Pathological Lives

RGS-IBG Book Series

For further information about the series and a full list of published and forthcoming titles please visit www.rgsbookseries.com

Published

Pathological Lives: Disease, Space and Biopolitics
Steve Hinchliffe, Nick Bingham, John Allen and Simon Carter

Smoking Geographies: Space, Place and Tobacco
Ross Barnett, Graham Moon, Jamie Pearce, Lee Thompson and Liz Twigg

Rehearsing the State: The Political Practices of the Tibetan Government-in-Exile
Fiona McConnell

Nothing Personal? Geographies of Governing and Activism in the British Asylum System
Nick Gill

Articulations of Capital: Global Production Networks and Regional Transformations
John Pickles and Adrian Smith, with Robert Begg, Milan Buček, Poli Roukova and Rudolf Pástor

Metropolitan Preoccupations: The Spatial Politics of Squatting in Berlin
Alexander Vasudevan

Everyday Peace? Politics, Citizenship and Muslim Lives in India
Philippa Williams

Assembling Export Markets: The Making and Unmaking of Global Food Connections in West Africa
Stefan Ouma

Africa's Information Revolution: Technical Regimes and Production Networks in South Africa and Tanzania
James T. Murphy and Pádraig Carmody

Origination: The Geographies of Brands and Branding
Andy Pike

In the Nature of Landscape: Cultural Geography on the Norfolk Broads
David Matless

Geopolitics and Expertise: Knowledge and Authority in European Diplomacy
Merje Kuus

Everyday Moral Economies: Food, Politics and Scale in Cuba
Marisa Wilson

Material Politics: Disputes Along the Pipeline
Andrew Barry

Fashioning Globalisation: New Zealand Design, Working Women and the Cultural Economy
Maureen Molloy and Wendy Larner

Working Lives – Gender, Migration and Employment in Britain, 1945–2007
Linda McDowell

Dunes: Dynamics, Morphology and Geological History
Andrew Warren

Spatial Politics: Essays for Doreen Massey
Edited by David Featherstone and Joe Painter

The Improvised State: Sovereignty, Performance and Agency in Dayton Bosnia
Alex Jeffrey

Learning the City: Knowledge and Translocal Assemblage
Colin McFarlane

Aerial Life: Spaces, Mobilities, Affects
Peter Adey

Millionaire Migrants: Trans-Pacific Life Lines
David Ley

State, Science and the Skies: Governmentalities of the British Atmosphere
Mark Whitehead

Complex Locations: Women's Geographical Work in the UK 1850–1970
Avril Maddrell

Value Chain Struggles: Institutions and Governance in the Plantation Districts of South India
Jeff Neilson and Bill Pritchard

Queer Visibilities: Space, Identity and Interaction in Cape Town
Andrew Tucker

Arsenic Pollution: A Global Synthesis
Peter Ravenscroft, Hugh Brammer and Keith Richards

Resistance, Space and Political Identities: The Making of Counter-Global Networks
David Featherstone

Mental Health and Social Space: Towards Inclusionary Geographies?
Hester Parr

Climate and Society in Colonial Mexico: A Study in Vulnerability
Georgina H. Endfield

Geochemical Sediments and Landscapes
Edited by David J. Nash and Sue J. McLaren

Driving Spaces: A Cultural-Historical Geography of England's M1 Motorway
Peter Merriman

Badlands of the Republic: Space, Politics and Urban Policy
Mustafa Dikeç

Geomorphology of Upland Peat: Erosion, Form and Landscape Change
Martin Evans and Jeff Warburton

Spaces of Colonialism: Delhi's Urban Governmentalities
Stephen Legg

People/States/Territories
Rhys Jones

Publics and the City
Kurt Iveson

After the Three Italies: Wealth, Inequality and Industrial Change
Mick Dunford and Lidia Greco

Putting Workfare in Place
Peter Sunley, Ron Martin and Corinne Nativel

Domicile and Diaspora
Alison Blunt

Geographies and Moralities
Edited by Roger Lee and David M. Smith

Military Geographies
Rachel Woodward

A New Deal for Transport?
Edited by Iain Docherty and Jon Shaw

Geographies of British Modernity
Edited by David Gilbert, David Matless and Brian Short

Lost Geographies of Power
John Allen

Globalizing South China
Carolyn L. Cartier

Geomorphological Processes and Landscape Change: Britain in the Last 1000 Years
Edited by David L. Higgitt and E. Mark Lee

Pathological Lives

Disease, Space and Biopolitics

Steve Hinchliffe, Nick Bingham,
John Allen and Simon Carter

WILEY Blackwell

This edition first published 2017
© 2017 John Wiley & Sons, Ltd

Registered Office
John Wiley & Sons, Ltd, The Atrium, Southern Gate, Chichester, West Sussex, PO19 8SQ, UK

Editorial Offices
350 Main Street, Malden, MA 02148-5020, USA
9600 Garsington Road, Oxford, OX4 2DQ, UK
The Atrium, Southern Gate, Chichester, West Sussex, PO19 8SQ, UK

For details of our global editorial offices, for customer services, and for information about how
to apply for permission to reuse the copyright material in this book please see our website at
www.wiley.com/wiley-blackwell.

The right of Steve Hinchliffe, Nick Bingham, John Allen and Simon Carter to be identified as the
authors of this work has been asserted in accordance with the UK Copyright, Designs and Patents
Act 1988.

Library of Congress Cataloging-in-Publication Data

Names: Hinchliffe, Steve, 1967– author. | Bingham, Nick, author. | Allen, John, 1951– author. |
  Carter, Simon, 1960– author.
Title: Pathological lives : disease, space and biopolitics / Steve Hinchliffe, Nick Bingham,
  John Allen, and Simon Carter.
Description: Chichester, West Sussex ; Malden, MA : John Wiley & Sons Inc., 2017. |
  Includes bibliographical references and index.
Identifiers: LCCN 2016028691 | ISBN 9781118997598 (hardback) | ISBN 9781118997604
  (paperback) | ISBN 9781118997628 (epub) | ISBN 9781118997611 (Adobe PDF)
Subjects: LCSH: Microbial ecology–Health aspects. | Bioethics. | Communicable diseases–
  Epidemiology. | Human-animal relationships–Political aspects.
Classification: LCC QR100 .H56 2017 | DDC 577.8–dc23
LC record available at https://lccn.loc.gov/2016028691

A catalogue record for this book is available from the British Library.

Cover image: Photograph © Henry Buller, 2015

Set in 10/12pt Plantin by SPi Global, Pondicherry, India

Printed and bound in Malaysia by Vivar Printing Sdn Bhd

10 9 8 7 6 5 4 3 2 1

To the memory of Professor Doreen Massey
An infectious intellectual, colleague and friend

Contents

List of Figures

Series Editors' Preface

The RGS-IBG Book Series only publishes work of the highest international standing. Its emphasis is on distinctive new developments in human and physical geography, although it is also open to contributions from cognate disciplines whose interests overlap with those of geographers. The Series places strong emphasis on theoretically-informed and empirically-strong texts. Reflecting the vibrant and diverse theoretical and empirical agendas that characterise the contemporary discipline, contributions are expected to inform, challenge and stimulate the reader. Overall, the RGS-IBG Book Series seeks to promote scholarly publications that leave an intellectual mark and change the way readers think about particular issues, methods or theories.

For details on how to submit a proposal please visit:
www.rgsbookseries.com

David Featherstone
University of Glasgow, UK

Tim Allott
University of Manchester, UK

RGS-IBG Book Series Editors

Acknowledgements

The work for this book was made possible with the generous support of the UK's Economic and Social Research Council. Two awards, 'Biosecurity Borderlands' (RES-062-23-1882) and 'Contagion' (ES/L003112/1) allowed us to conduct the field investigations and develop key conversations with actors in the field. The work has also benefitted from our co-researchers in those projects, especially the contributions of Dr Stephanie Lavau and Dr Kim Ward. Their observations and work on this project while at Exeter University have been instrumental in the generation of this book.

At the UK's Animal and Plant Health Agency, Steve Hinchliffe would like to thank Drs Richard Irvine, Jill Banks, Andrew Breed and Professor Ian Brown who were all incredibly patient as he learned the language of influenza viruses and animal surveillance. At the Food Standards Agency, Steve Hinchliffe would like to thank the secretariat and colleagues in the Social Science Secretariat and Research Committee, especially Sian Thomas and Helen Atkinson.

At Exeter University we are indebted to conversations with a wide group of scholars – Michael Schillmeier, Henry Buller, Katie Ledingham, Gail Davies, Ann Kelly, John Dupré, Astrid Schrader, Elizabeth Johnson, Jo Little, Robbie MacDonald, Sarah Crawley, Jamie McCauley, Andrew Pickering, Sabina Leonelli, Krithika Srinivasan, Sam Kinsley, Regenia Gagnier among many others.

A host of others have commented upon or shared conversations around this work. They include Carlo Caduff, Ian Scoones, Paul Forster, Melissa Leach, Gareth Enticott, Ben Fine, Susan Craddock, Melanie Rock, Bruce Braun, Frederic Keck, John Law, Annemarie Mol, Beth Greenhough, Jamie Lorimer, Andrew Barry, Pierre-Olivier Methot, Alex Nading, Rob Wallace, Andrew Donaldson, Kezia Barker, Stephan Price, Kristin Asdal, Linda Madsen and Ann Bruce.

We would also like to thank Dave Featherstone for his support and patience as editor of this series, and for the thoughtful comments of two anonymous referees whose comments were invaluable at the editing stage.

Chapters 4, 5 and 8 are substantial re-workings of material that is published in the *Journal of Cultural Economy*, *Geoforum* and in the edited book *Humans, Animals and Biopolitics: The More Than Human Condition* (Kristin Asdal, Tone Druglitrø and Steve Hinchliffe).

Foreword

Pandemics, epidemics, zoonoses and food-borne diseases have, for some at least, become key challenges for contemporary global society. They threaten progress in global health, compromise food security, and, along with climate change and global terrorism, seem to usher in a state of emergency and a radically uncertain future. Just as importantly, they are associated with often painful and life altering illnesses that can exact suffering on the part of people and animals as well as social and economic hardship.

Many of these diseases are associated with what have become known as emerging and re-emerging infections, a term that is often associated with a dynamic and unpredictable microbial world of recombinant viruses, resistant bacteria and mobile microbial genes. These microbes are testament, if any was needed, to the liveliness of the non-human world.

And yet these diseases are more than a matter of microbes alone. They are instead a product of relations that involve microbes, their hosts and their social as well as physical environments. That is, they are the result of bio-social clusters that involve and re-format economies, social practices, living bodies and microbes.

So rather than disease being 'out there' or 'to come', in the form of a pathogen waiting to attack, we prefer to think of life as to greater or lesser extents pathological, or prone to disease. For us, it is the configuration of various matters and living processes that makes life more or less healthy.

This is not to say that all lives are equally diseased, or that disease is everywhere, or that it is somehow already present. Clearly there are disease and illness events that mark distinct and often irreversible ruptures in daily life. But it is to say that we can usefully identify some lives and ways of living as more pathological than others. How these pathological lives fare is dependent, we argue in this book, on the quality of the spatial relations from which they are made.

So, rather than focus on pathogens and their exclusion from everyday living spaces as a means to address the threat of emerging disease, we take a different tack. We use pathological lives as a means to understand how so many contemporary human and non-human animal lives are living on a knife-edge. In this view, the apparent stability of modern lives may, paradoxically perhaps, exhibit a form of fragility that is borne from their being lived at a threshold. It may take little, in other words, to push them over the precipice and into a pathogenic state.

Keywords

The book addresses a puzzle of how best to understand and respond to the rise of interest in and concern over emerging infectious and food-borne diseases. What is contributing to the recent growth of disease threats? How can we account for their persistent presence on political and public health agendas? In order to answer these questions we have mobilised some key terms that we will briefly introduce here before they are developed in detail in later chapters. These terms are not fixed in stone, but we hope they provide a handhold through the book's chapters and to the issue of pathological lives.

The first key term is **pathogenicity**, a word we use to highlight that infectious disease is always more than a matter for pathogens alone. In its more scientific usage, the term refers to the potency or effectiveness of a particular pathogen (a bacteria, virus or other parasite for example). But here we mean to underline the relational ways in which infectious diseases are made. In the simplest of senses this can refer to the basic notion that diseases are made from host-pathogen and environmental interactions. Pathogenicity is in this understanding borne out of the kinds of relations that hosts have with bacteria and viruses, their vectors and so on. A healthy host within a healthy population and environment is likely, for example, to reduce the pathogenicity of a microbe. In conditions of vulnerability, however, an otherwise inconsequential infection can take on life-threatening qualities. In this book we supplement this epidemiological 'matter of fact' with other relations that contribute to pathogenicity. They include economic relations generated through markets for livestock and their produce, labour relations that format the interrelations between human and animal hosts, governmental relations that affect the ways in which diseases are monitored and so on. Our argument is that it is the intensities of these relations, their spatial interrelations, that constitute the pathogenicity of a disease. Pathogenicity, or the ability for diseases to amplify and reach new levels of intensity, is an outcome of these and many other spatial and socio-material relations.

The second key term is **disease diagrams**. Diagrams refer to the ways in which diseases are understood and acted upon. For example, an infectious disease may be diagrammed as something to keep out through the installation of a barrier. Or it may be something to intervene in through the production of a vaccine, or other medicine. As might be obvious, there is often more than one of these diagrams in play at any one time, and the relative emphasis given to one diagram, or the specific mix of diagrams in play, can have effects on how a disease evolves, how it interrelates with other diseases, where authority lies in relation to disease and who is deemed as responsible for health. How people respond to this diagramming of disease becomes a critical issue for disease management.

The third key term is **disease situations**. If we take the socio-material intensities that generate pathogenicity along with the specific suite of disease diagrams that are mobilised to both understand and intervene in disease dynamics, then together they start to define what we are calling a disease situation. In using the term, we have the following intentions:

- Situations are, first of all, meeting places, where numerous actors, bodies, species, pressures, flows, issues, decisions and so on are organised or brought together, or held apart or worked upon. They are heterogeneous (formed from their differences and relations), more than human, and dependent not only on what is meeting up but also how those meetings are spatially configured.
- Situations, like pathogenicities, are relational – that is their properties or character are generated by and generative of social, spatial and material relations. They are not structures. They are grounded in practices and orderings and are as such more or less open to change. Situations clearly owe a debt to the relational geographies that precede this work (Murdoch, 2005; Whatmore, 1997) as well as to a more general interest in spatial analysis and thinking topologically.
- Situations bear a family resemblance to the notion of assemblage, or the inter-relations and co-production of various 'species' or unlike kinds (diagrams, microbes, populations and so on). Assemblage in our view is a process, and differs from a whole or a system in that these unlike kinds need have nothing in common. No one in that sense can speak for the whole situation. Nevertheless, these 'species' inter-mingle and can radically affect one another within their situation.
- Finally, and in a way that takes us beyond some treatments of assemblage, situations are more than descriptions of the atmospheres generated by the convergence and divergence of various interrelating practices and matters. They are also, crucially, eventful and as such may offer the ingredients for change and intervention. When you are in a situation you are invited to act. In other words, situations have a potentiality that can generate events, prompt a shift in attention and foment new actions. Situations are in that sense real and existing manifestations of multiple processes that also have a power to force thought. To be clear, this power is always 'a virtual one' that 'has to be actualised' (Stengers, 2005b: 185). How this power can be realised (by, for example, those who attend to the more than human details of a situation) is a matter that is taken up in later chapters in the book.

Together these terms help us to offer original insights into the current disease predicament. Once we take a situational approach, with its pathogenicities and diagrams, we can identify the ways in which many current approaches to infectious and food-borne diseases tend to miss some vital clues in terms of how to make life safe. Or, worse, how these same approaches may in fact, and paradoxically, make life even less safe. Our contention is that once we take disease situations seriously we can start to question the norms and assumptions that so often under-line current re-investments in life politics. Our argument is that we need to move away from a version of life politics (bio-politics) where norms are policed and re-enforced (often materially with a system of barriers and boundaries) to a lively politics (cosmopolitics) where we can use current disease situations to start to trace counter-norms, to identify suppressed modes of existence and in doing so find possible ways out of the current predicament.

Reading Pathological Lives

In order to develop these arguments, we have divided the book into two main sections. In Part I (Chapters 1 to 3) we introduce the book's approach and expand on the conceptual and methodological issues that relate to emergency diseases and pathological lives. We start by asking how emerging infections and food-borne diseases have been conceptualised or framed, and how these have informed approaches to disease management. Working spatially, in Chapter 2 we adopt the term 'disease diagram' to chart the history of approaches to and interventions in infectious disease and follow this with an account of how and why current approaches to disease tend to involve a particular mix of disease diagrams. The particular mix of diagrams is, we argue, both a matter for empirical enquiry and a key feature of what we call a disease situation. In Chapter 3 our attention becomes more methodological as we expand on what we mean by 'disease situa-tions' and ask how we might re-configure conventional approaches to infectious disease. We outline some of the shifts required as we move from a geometry or topography of disease, with its focus on disease spread or extension over space, to one that is more attendant to the topologies of disease situations. The latter is concerned with the spatial intensities and relations that are generated in particular set ups and that make disease more or less likely.

Having set up diagrams and situations in Part I, in Part II we focus on a range of disease situations, or specific cuts through those situations. Following a short introduction to Part II, we start, in Chapter 4, in the hen house and look at the poultry industry as a main player in the re-diagramming of avian and zoonotic diseases. We then move, in Chapter 5, to the pig sty, and chart not only the pres-sures that make a disease situation but also the efforts by farmers and others to patch and piece together healthy lives. Food-borne diseases are our concern in Chapter 6, as we leave the farm and look at the ways in which food chains are understood and regulated. If Chapters 5 and 6 introduce the fraught politics of attention into pathological lives, Chapter 7 asks how likely this different politics of life might fare in what we call disease publics. Chapter 8 uses fieldwork on wild and domestic birds, and on viruses, to look for a different kind of life politics that may be developed from these situated knowledges. Finally, in the Conclusions, we spell out what a re-oriented and spatialised politics of life means for the infectious disease issue.

It is of course possible to read the chapters individually, and to move between the situations that we trace in Part II, but they are not entirely stand-alone. Our hope is that the narrative of the book will carry readers through the various arguments developed in Part I, to the more empirical treatments in Part II. Note however that we rarely treat the theoretical and empirical as distinct. The arguments in Part I are all empirically grounded, while the situations in Part II often involve conceptual development.

Part I
Framing Pathological Lives

Chapter One
Pathological Lives – Disease, Space and Biopolitics

The diversity and geographical distribution of influenza viruses currently circulating in wild and domestic birds are unprecedented since the advent of modern tools for virus detection and characterisation. The world needs to be concerned (WHO, 2015).

The number one risk on the [UK] Government's national risk assessment for civil emergencies, ahead of both coastal flooding and a major terrorist incident, is the risk of pandemic influenza (House of Commons Committee of Public Accounts, 2013: 6).

We are left in the hands of the generations which, having heard of microbes much as St Thomas Aquinas heard of angels, suddenly concluded that the whole art of healing should be summed up in the formula: Find the microbe and kill it. And even that they did not know how to do (George Bernard Shaw, 1909: Preface to the *Doctor's Dilemma*).

Introduction: The Emergency of Emergent Infectious Diseases

In Western states, at least, emerging infectious diseases have become emergencies in waiting. The threat of a widespread and acute malady affecting people, or indeed the plants and animals on which they rely (to say nothing of the technical infrastructures or other living and non-living networks which sustain

Pathological Lives: Disease, Space and Biopolitics, First Edition. Steve Hinchliffe, Nick Bingham, John Allen and Simon Carter.
© 2017 John Wiley & Sons, Ltd. Published 2017 by John Wiley & Sons, Ltd.

life), is a long-standing one. But it is the imagined and to some extent experienced interdependencies and vulnerabilities that people share with each other and with other living bodies that seem to have raised the stakes in the last few decades. So much so that we are, for some at least, 'teetering on the edge' of a major disease event or catastrophe (Webster and Walker, 2003).

Simplifying somewhat, this apparent emergency-to-come has two core elements. First, it is *socio-ecological* and based on the sciences of 'emerging and re-emerging diseases'. Here the focus tends to be on mutable microorganisms and the potential for those organisms to wreak havoc in a highly 'infectable' and densely interconnected modern world (Braun, 2007). Previously, and in the main, microorganisms had been understood as more or less fixed entities that would inevitably run their evolutionary course, becoming less significant over time (Methot and Fantini, 2014). The emergence of new infectious diseases, like AIDS (Acquired Immune Deficiency Syndrome) and SARS (Severe Acute Respiratory Syndrome), and the re-emergence of newly virulent scourges (like influenza and tuberculosis) in the later part of the twentieth century suggested that life, and microbial life in particular, was less fixed and so less predictable than we might have countenanced.

These new agents of concern could mutate and recombine, jump species, take advantage of new environmental conditions, and could move through the dense and rapid transit routes that circled the planet. Instead of being on the wane, microbes were back on the agenda. Along with climate change and global terrorism, they formed a raft of 'agents' that were regarded as mutable, indeterminate and generative of catastrophic events. For some at least, global connectivity and molecular mutability had combined to usher in a new age of plagues, epidemics and pandemics (Garret, 1994). Arguably, emerging diseases suggested that the world was now more 'infectable' than ever.

Second, this emergency-to-come is *governmental*. Here, emergency relates to a form of anticipatory or future-oriented government that seeks to highlight (even give greater emphasis to) potential breakdowns in social order. In this style of governance of and through emergencies – which has arguably become dominant in recent decades (Amoore, 2013) – the role of public and private institutions is to organise for events that are of sufficient magnitude that they demand foresight and preparation (Anderson, 2010; Collier and Lakoff, 2008b).

Pandemics, infectious animal diseases and food contamination events, for example, can all exact such far-reaching challenges to social and economic life that they constitute security issues, necessitating some kind of civil emergency planning in order to prepare for or mitigate their worst effects. Infectious disease, in this sense, has become part and parcel of a logic and practice of security. Indeed, the term biosecurity is often used in relation to the threat of emerging infectious diseases, and refers to the raft of measures and policies that governments, commercial and other organisations seek to put in place in order to reduce the risk of a disease event and/or prepare for the consequences of such an event in terms of emergency response. Whether the resulting biosecurity practices are effective or make matters more prone to go wrong is a major question that we will return to throughout this book.

These socio-ecological and governmental aspects of emerging infectious diseases may well be mutually re-affirming. First, and most obviously, changes to the 'infectability' of the planet may be accompanied by shifts in approaches to infectious disease control, leading to the rise of biosecurity as a discourse and material practice. Second, the rise of a form of anticipatory governance clearly requires its own set of justifications. Mutable microbes form a convenient ontology or cause under which new kinds of human authority and control can be justified and normalised (King, 2002). For some commentators, a mode of life (a *modus vivendi* (Sloterdijk, 2013)) emerges in which control is predicated on accentuating certain threats. This may be more than ideological. Perhaps what is most interesting here is the possibility that, third, these forms of human control can in turn produce new microbial environments that may inadvertently be even more challenging. Spiralling efforts to counter microbial threats, or a hypertrophic approach to security, can seed further changes to rapidly evolving microbiomes (Landecker, 2015). So much so that a belief in human authority and control may well be part of the problem.

Pathological Lives engages with the disease emergency, its rise up scientific and political agenda, its formatting through biosecurity and, crucially, the extent to which the resulting foci of attention may well be making matters worse rather than better. We focus on the particular ways in which emergency diseases are constituted – how they are understood, marshalled, measured, generated and even ignored. Our method is at once geographical and based within science and technology studies. In taking these approaches, with their legacy of fieldwork and 'theorising empirically' (Mol, 2002), we are interested in the practical 'doings' of disease rather than the grand stories that are told about them. Only by investigating practices (what is done as well as what is said) can we assess the extent to which these doings may play a part in bringing about the emergency they seek to mitigate – or, indeed, may offer new openings for doing things otherwise.

This book is empirically grounded, and in being so it can make some claim to a better understanding of how communicable diseases are being managed and mismanaged and will aim to make some concrete suggestions about what it takes to do health and disease in ways that are better suited to the current predicament. It is based on fieldwork that we undertook across a range of sites and involving all manner of species (from farms to restaurants, from wildlife reserves to virological laboratories, and from factories to living rooms). Our methods are varied, though mostly ethnographic in character and sensibility, and our aim has been to allow the practices that we have observed, written down and questioned, to surprise us, to put our concepts at risk and to force us to think carefully and critically about the disease emergency.

Pathological Lives is also conceptual. In working across numerous sites and species, and in linking together those sites and species, our argument pulls together key geographical or spatial insights on the relations between people, animals, microbes, infrastructures and ways of governing disease. Rather than focusing on one part of the disease system, we are interested in what happens when you take the changing relations between hosts, microbes and their environments, as well as the emerging regimes of control or governance, as the key concern for investigation. Infectious disease in this instance, and for us, is not only a result of microorganisms infecting a host, but the multi-faceted outcome of the changing relations that make microbes more or less likely to be effective in generating disease.

The key question becomes, in this sense, how various matters (including not only microbes) combine with other conditions to produce disease. We make a distinction, then, between those approaches that focus on disease as a matter of discrete causative agents and those that view disease as a more relational phenomenon. In the first instance, approaches to infectious disease management or control that focus on microbes as pathogens tend to emphasise their absence and exclusion. They involve constructing and maintaining real as well as metaphorical walls (see Figure 1.1).

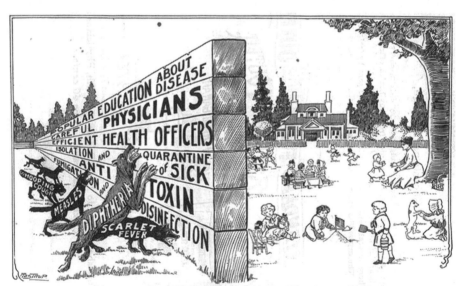

How High is the Wall in Your Town?

Figure 1.1   How safe is your town? – US Public Health Poster, undated. The divisions between a contingent wild world and a domestic culture are clearly marked as spatially exclusive. The existence of a single non-human on the right side of the wall, on two legs and with tail docked, is emblematic of this nature/culture binary that is in play. (*Virginia Health Bulletin*, 1908: 216).

This so-called 'ontological' approach to disease, with microbiological organisms as agents of causation, is often in tension with the second, relational or more 'physiological' and 'ecological' understandings of disease (Anderson and MacKay, 2014). In the latter, microbes may be just one component of a complex of matters that conspire to produce disease. Here, understanding disease is less likely to focus on a single agent and pathway to the host, but on a suite of issues, on biographical details (a patient or host's propensity to develop symptoms) and on the patient's social and ecological setting (relating to a population, its density, levels of immunity and so on). It may even stretch to consider the role of disease management and governance in the shaping of the disease.

In these ecological approaches, the focus may be less concerned with keeping matters out and more attentive to the multiple relations that make disease. How these relations are configured, spatially, becomes the key matter of concern. It is the constellation of matters that twist and turn bodies of all shapes and sizes into diseased relations that grab our attention. All life in that sense is more or less pathological. It is the quality of the relations that make those lives more or less liveable.

In order to make these spatial arguments we employ some key terms. First, in Chapter 2, we introduce disease diagrams, which we take to be ways of understanding and acting upon disease and health threats. For example, if disease is understood as a germ-borne menace, then exclusion is the spatial practice of choice. Conversely, if a more ecological or biographical approach is taken, then disease may be diagrammed as a matter of social inclusion and public health campaigns (perhaps by improving health services, availability of vaccines and so on). Diagrams then are the ways that disease is grasped and governed. They are spatialisations that affect the ways in which disease and health are understood and treated. For sure, these and other disease diagrams may co-exist and be in tension with one another. How one or other gains ascendancy or how they are mobilised at the same time and in the same place becomes a key resource for the interrogation of disease in practice.

The second term we adopt is disease situation (Chapter 3). This refers to the ways in which a specific combination of disease diagrams, as well as the suite of issues that make disease more or less likely (including, for example, host population characteristics, forms of governance, market pressures and so on), generate the conditions for living. This, in short, is an ecology or assemblage – a spatial arrangement or meeting of ideas, practices and materials. But more than this, a situation also alerts us to the possibility that this configuration or assemblage has a 'virtual power' to force thought, to make us think again about disease and health (Stengers, 2005b: 185). This power has to be realised, and one of the jobs of analysts is to help to bring that forcing of thought into being.

To be clear, in using the term ecology in this context we do not mean to signal a science of functions, or simply an adding together of 'factors'. The components of a situation will combine together in ways that are not easy to predict and cannot be assumed to be simple additions. This is because, in the terms of Karen Barad (2007), matters will intra-act rather than simply interact. That is, they are relational and may alter each other as they go. So, the situations we describe are close to what Isabelle Stengers has described as an ecology as 'a science of multiplicities, disparate causalities, and unintentional creations of meaning' (Stengers, 2010a: 34). This space of crossings, folds and missed opportunities may, we argue, open up new ways of imagining and enacting the politics of disease and the definition of what counts as the disease emergency. It may, we suggest, help to empower the disease situation by redefining what counts as the emergency of emerging infections.

These terms and the approach to disease situations as possible sources for redoing pathological lives course through the book in its conceptual framing (Part I), its more empirically focused chapters (Part II), and in the conclusion. To spell out some of these possibilities and the approaches on which they are based, we next draw out four key moves that distinguish the approach that we take to the spatial politics of disease. ˙

The Four Moves of Pathological Lives

There are four key moves that we make in this book that distinguish our approach to emergency disease. First, we revisit the emergent infectious disease thesis and justify the book's shift in focus *from forest edge to socio-technical diseases*. Infectious disease becomes, on this account, a networked matter involving markets, sciences, governments and so on. Second, we note how this refocusing of attention on socio-technical set-ups requires a rather different spatial imagination, as we move *from disease sites to disease situations*. Third, *from pathogens to pathogenicities*, we add to this spatial reconfiguration by expanding on a distinction between a pathogen-focused understanding of disease and one that is more interested in diseases as relationally produced. Finally, in *a politics of life*, we briefly discuss biopolitics, one of the key means through which the management of emergency diseases has been organised and conceptualised, and in doing so open up a conceptual frame that provides some resources for generating a new or different politics of life.

From Forest Edge to Socio-Technical Diseases

In the early 1990s, communicable and infectious diseases were back on the political agenda. They had been somewhat marginalised following the triumphant post-war pronouncements that the world was on the brink of an epidemiological transition. Communicable diseases would be relegated, so the optimists argued, to a relatively minor component of human morbidity and mortality through the use of antimicrobial medicines, improved hygiene and other modern technologies (Omran, 1971). This technological and medical optimism was 'flanked by a belief in the natural decline of virulence' (Methot and Fantini, 2014: 218), or the gradual co-evolution of microorganisms and their hosts resulting in better adaptations or a shift from pathogenic to commensal relations. Yet, as the century neared its end, a raft of diseases started to unsettle the progressive narrative of continuous medical advance and infectious disease decline. Some of these diseases were new, while others had been persistent scourges for people in the Global South and were now threatening to infect the North (Farmer, 1999).

These new and neglected infectious diseases owed some of their rise to prominence in the Global North to a renewed fear that emerging infectious diseases had the potential to produce global pandemics. The concerns seemed to restage a rather ancient fear of contagion and connectivity in which certain parts of the world, certain social groups and certain practices are pathologised and demonised (King, 2002). The 'virtuous' Global North and the under-regulated and therefore contingent Global South seemed, in disease terms, closer than ever. The alarming rhetoric of cosmopolitan life out of control, common in international organisations (World Health Organisation, 2007) and popular science writing (Garret, 1994; Garrett, 2013; Quamenn, 2012; Wolfe, 2011) made use of familiar spatial imaginaries where 'modern' society is threatened by poorly governed, exotic, other-worldly, sexualised and often naturalised lives. The enemy figures of mutating, slippery, re-assorting pathogens, of free-roaming super-spreaders and patient zeros, of (largely non-Western) human-animal practices or interspecies intimacy (Shukin, 2009: 46) and of rogue practitioners in an otherwise 'orderly' system, are all portrayed as 'outsiders' in need of excommunication.

As a result, an 'Out of Africa' and or 'Out of Asia' mapping of emerging infections is common (Leach and Dry, 2010). Indeed, the emergence of the Human Immunodeficiency Virus (HIV) in the late 1970s was followed by the identification of an infective pathway that implicated initial human contact with non-human primates. Forest edges within central African states, and the hunting and consumption of bush meat, became key sites for virological research (Wolfe et al., 2007). Simian immunodeficiency viruses (SIVs) and Ebola Haemorrhagic Fever (EHF) provided similar geographies of emergence, with unregulated primate contact as the disease transfer event or first cause. Moreover, other diseases, including Japanese Encephalites, Nipah and Hendra, often with bats as wild hosts, tended to draw researchers to the newly opening borderlands between wildlife and human society.

These emergence and transmission narratives repeated a familiar story of human encroachment, ecological change and intensifying interactions with and/ or disturbances to wildlife that together produced the conditions of possibility for the transfer of microorganisms from wildlife to people. While not new, these transfers now had more chance of circulating beyond the confines of the forest edge as a result of new transport infrastructures that were often associated with military conflict, science, intensive forestry, mining or agriculture. Stephen Morse, the virologist who did much to popularise this etiology, coined the term 'viral traffic' as a means to capture the direction, increased reach and accelerating speed of viral movement (Morse, 1993). He also coined the terms 'emerging infectious diseases' and 'emerging viruses' to relay the sense of the effects of a combination of human-induced environmental changes, shifts in human–non-human animal interactions and increased transport and communications on the evolution and disease-producing capacity of microbial life.

The promiscuity of human and non-human lives, their mixing, movement and co-dependencies, seemed to drive a continuous 'spill-over' (Quamenn, 2012), where microbes that were once restricted to non-human species were able to transmute and transmit to people. The evidence was arresting. Roughly three-quarters of the emerging diseases of the last three decades were judged to be 'zoo-notic' (infectious diseases that jump between human and non-human animals (Taylor et al., 2001)) and 60% of all known human communicable diseases were 'due to multi-host pathogens characterised by their movement across species lines' (AVMA, 2008: 3). At the turn of the new century, a wave of zoonotic respiratory diseases increased the stakes further (at least in the West and Global North), giving the emergency in waiting further credence. Severe Acute and Middle East Respiratory Syndromes (SARS and MERS) in 2002 and 2013, as well as Avian and Swine Influenzas throughout the 2000s and in 2009, seemed to testify to this upturn in infectious and often viral diseases that had multiple non-human hosts and or vectors and were associated with high mortality rates in people.

In terms of the geographies of infectious disease, these diseases started to shift the epidemiological gaze away from the forest edge. SARS seemed to be more urban and peri-urban in terms of its epicentre (Harris Ali and Keil, 2008; Schillmeier, 2013). The new wave of zoonotic influenzas like avian and swine flu were clearly related to socio-economic conditions and even to industrial practices (Wallace, 2009). Certainly, the geography of emergence and spread outwards from a hot spot in the Global South was always questionable to social scientists interested in the relations that make disease possible, but these epidemic and pandemic events made this geography even less convincing. The 2009 swine flu pandemic strain virus seemed, for example, to emerge within intensive pig raising facilities in North America and Mexico. Meanwhile, highly pathogenic avian influenza (HPAI) viruses seem to have both wild and domestic birds to thank for their emergence and continuing evolution. The focus of attention has therefore somewhat shifted from forest edges and wildlife to semi-domesticated and domesticated non-human animal hosts that are more centrally linked to food and farming practices as well as to the laboratory and regulatory practices that are associated with securing safe life.

This refocusing attention on the socio-technical disease set-up is further justified when we consider other disease emergencies. For alongside zoonotic infections, there are also diseases that affect often large and vulnerable domestic animal populations, with devastating effects on animals, economies and the people who work with those animals. The disease pathways and mechanisms may be similar, often with wildlife hosts acting as 'reservoirs' for microorganisms and playing a key role in the cycling and recycling of disease. However, of equal concern here are the growing size and scale of domestic animal populations and holdings. In the last few decades the rise of both global livestock animal numbers, particularly for chicken, cattle and pigs, and a continuing growth in average carcass weights, amount to a ballooning global domestic animal biomass (see Figure 1.2). In terms of disease risk, this expansion of mass increases the magnitude, if not the frequency, of the risks of epizootic events, or widespread non-human animal diseases. The implications for animal welfare, food security and livelihoods are clear. Again, recent disease events that have affected national herds like Bovine Spongiform Encephalopathy (BSE), foot and mouth disease, Bovine Tuberculosis, Brucellosis, Porcine epidemic diarrhoea and others, are indicative of the kinds of vulnerabilities that these livestock operations face.

To add two more concerns, there are also food-borne diseases and the emergence of anti-microbial resistance. Food-borne diseases can develop within living animals or on animal products that support a diversity of life, and travel with those products through a convoluted pathway and complex food industry to reach consumers in forms that are difficult to monitor and control. The food industry itself becomes the site of emergence, so much so that microbiological life can flourish and mutate within its fabric and generate new forms of disease. BSE, *Campylobacter* (a bacteria associated with food poisoning) and *Escherichia Coli* 0157, all seem to have opportunistically combined rapid microbiological change with a complex and high-pressure food chain to produce new challenges to public health.

Finally, there is the spectre of anti-microbial resistance (AMR). The rise of antibiotic treatments has enabled highly successful control of bacterial infections in humans and non-human animals. But resistance to the effects of these naturally occurring and synthetic medicines is an inevitable part of microbial evolution. This process is accelerated by the misuse or poor stewardship of those medicines. The result is that there are now strains of bacteria associated with food and farming (*E. Coli*, Methicillin-resistant *Staphylococcus aureus* (MRSA) and *Campylobacter*) that are resistant to first-line treatments and broad-spectrum antibiotics. The circulation of resistant microbes and their mobile genes through the food and farming system and into the wider environment via human and animal wastes is a key concern (Wellington et al., 2013).

Taken together, these zoonotic, epizootic, food-borne and AMR emergencies not only shift attention from the forest edges towards socio-technical disease setups, they also suggest a shift in the role of social science (Janes et al., 2012). To put it simply, these set-ups involve a broader set of relations with non-human animals than might be found at the forest edge. The issue may no longer be the encroachment of people into wild spaces, but quite the reverse, the increasingly

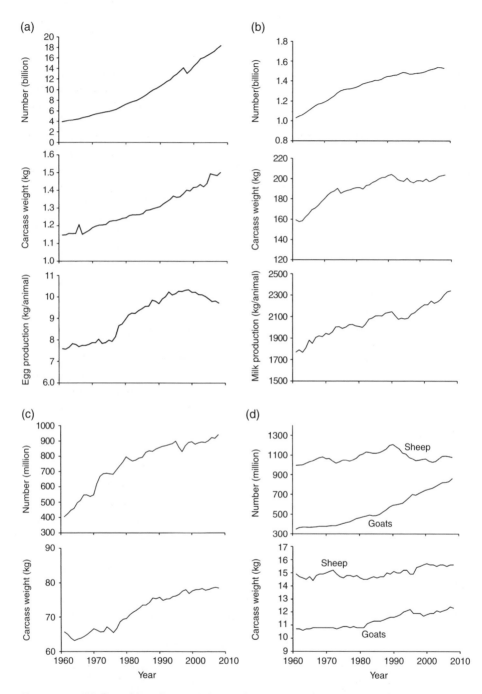

Figure 1.2 Global livestock production: (a) number of chickens, carcass weight and egg production per animal from 1961 to 2008, global; (b) number of bovines (cattle and buffaloes), carcass weight and cattle milk production per animal from 1961 to 2008, global; (c) number of pigs and carcass weight from 1961 to 2008, global; (d) number of sheep, goats and carcass weights from 1961 to 2008, global. (Thornton, 2010, http://rstb.royalsocietypublishing. org/content/365/1554/2853. Used under CC BY 4.0 https://creativecommons.org/licenses/by/4.0/).

obvious encroachment of non-human animal ecologies *on* people (Nading, 2013). In other words, there is a pressing need to focus on the effects of an expanding domestic animal ecology, and to give more attention to the human-animal and material ecologies that are being reformatted in current iterations of agriculture, food provision, regulation and science.

As a result we primarily deal here with the socio-technical aspects of infectious disease as they relate to food and farming, where the generation, amplification and subsequent transmission of disease can produce catastrophic effects, and where the control of disease risk can, we argue, make matters both better and worse. In making this our focus we nevertheless take a multi-sited approach, one that includes farms, laboratories, factories, wildlife reserves, restaurants and kitchens, government offices and public meetings. How to tie these various sites together requires us to say a little more about what we call disease situations.

From Disease Sites to Disease Situations

The responses of food, farming and policy sectors as well as scientific research to emerging infectious and food-borne disease are of key concern for future global health. Crucially, these responses are coloured by the operation of a raft of other pressures on food and farming. These include economic, environmental and regulatory pressures. We will run through some examples here before outlining how and why this matters for studying disease.

Responding to emerging diseases takes place against a backdrop of growing demands on the food and farming sectors to produce food for more people, at prices and with margins that are agreeable to consumers and producers. The production of animal protein is a huge growth area for many economies, and this growth is driving and responding to changing diets and lifestyle expectations. Increased livestock productivity is part and parcel of the new but somewhat familiar aims of sustainable intensification (Garnett and Godfray, 2012). Domestic livestock populations are estimated to have grown at around 2.4% per annum, with carcass weight per animal increasing year on year (Alexandratos and Bruinsma, 2012). Globally, 52 billion chickens are consumed annually (FAO, 2013). Worldwide pig production is struggling to match growing demand for pork in Asia in particular. Food networks are increasingly international and global financiers are investing in concentrated animal feeding operations (CAFOs) and livestock infrastructure on an unprecedented scale (Pew Commission, 2008; Wallace, 2009) as a means to meet this demand and of course to secure steady returns on investment.

This expansion of agri-food has numerous benefits, it is often argued, in terms of providing affordable food for a growing and developing human population in ways that might be expected to meet clear standards of production (McCloskey et al., 2014). Yet, it also faces large challenges, including ones that relate to environmental externalities (not least greenhouse gas emissions, the throughput of finite resources and nitrate pollution) and animal welfare concerns. In terms of health and disease, the issues are not straightforward. On the one hand, there are arguments that increased concentration and enclosure (or modernisation) increases the ability of the food and farming system to exclude pathogenic materials. On the other, logistical and health challenges of a densely interconnected system look ever more difficult and prone to catastrophic failure. As global animal mass grows, as wildlife is simultaneously displaced and competes with agriculture for habitat, and as food systems increase in terms of network length and connectivity, the potential for devastating zoonotic, animal and food related diseases may well have grown. All of these concerns bear down on an industry that is increasingly under pressure to produce plentiful food, at low cost and in ways that are environmentally sustainable and healthy. The result is that there are often tensions that exist between safety and profits, biosecurity and food security, biodiversity and risk, consumer demands for affordable goods and food safety.

Expansion and internationalisation of the food and farming system also poses governance challenges at a time of widespread crises in terms of resourcing, organising and generating and maintaining trust in public and private institutions. Internationalisation may well be accompanied by a reduction rather than rising of standards as conglomerates and corporate bodies aim to improve margins by taking advantage of low-cost labour, cheaper inputs and less stringent regulation. At the same time, a general tendency to reduce the 'regulatory burden' on businesses and offset the public cost of dealing with disease events has led to a 'neo-liberal' style reorganisation of state and local state regulatory infrastructures (from veterinary services to food inspection, and from government-led science to national health provision) (Maye et al., 2012).

More specifically, there have been attempts to redistribute responsibility onto private actors so that they take charge of their disease risk. Farmers for example, in the UK and New Zealand, have been encouraged to form (albeit highly subsidised) limited liability companies in order to control Bovine Tuberculosis through culling of wildlife. There have even been attempts to make compensation for disease breakdowns dependent upon farmers having taken necessary steps to improve the biosecurity of their premises (Donaldson, 2008; Mason, 2014). In the US, following the avian influenza outbreaks of 2015, farmers are required to meet certain biosecurity standards in order to be eligible for compensation following future outbreaks. Meanwhile, under the same rubric of neoliberal approaches to governing, there are attempts to develop market opportunities in health and disease abatement.

This is an area where disease (as emergency in waiting) chimes with the prospect and marketing of scientific innovation and pharmaceuticals. The always inevitable and imminent emergency is a useful means to implore governments and businesses to invest in security, in frontline broad spectrum drugs, in disease resistant genetically modified or edited animals and so on. The market opportunity of security adds another pressure onto the mix of issues that surround disease management and control.

The broader point is that the intensification and extension of international food supply chains has been accompanied by re-arrangements to public budgets, an increase in private health and security providers, a reformatting of animal and human health related sciences, and, arguably, an atmosphere of scepticism and mistrust of both public and private authorities.

Given these material, social, ecological and political tensions, how should emergency diseases be studied? We are interested in the interfaces between the regulation or management of disease and other concerns (costs, labour dynamics, food safety concerns, ecology, issues of countryside and wildlife), an interest that requires us to broaden the spatial vocabulary through which diseases are normally studied. Here we make a distinction between a site (or a location on a map like a specific forest, or a farm, or a factory) and a situation in which various processes, diagrams, materials and actors of all shapes and sizes that make that place are brought into view (Chapter 3). If the forest edge has often been depicted as a disease site, or a location where a zoonotic event took place, then the situation that is a socio-technical disease set-up, or any moment within that set-up (like a farm, a processing plant, the decision to hire a certain kind of labour, a laboratory where viruses are identified, the ways in which regulation is organised) is always shaped by a any number of issues and practices, many of which occur elsewhere.

A situation is at once grounded somewhere but also dispersed or distributed through the many interactions that make it possible and which it can also affect. For us, situations are meeting places, where numerous actors, bodies, species, pressures, flows, issues, decisions and so on are organised or brought together, or held apart or worked upon. They are heterogeneous (formed from their differences and relations), and dependent for their identity not only on what is meeting up but also how those meetings are configured. The way in which these meetings occur, and the ways in which actors and materials intra-act (for they can and do change one another in the process of relating to one another, see (Barad, 2007)) is of key concern to a social science of disease.

The way labour practices intra-act with poultry guts (Chapter 4), or changing farming practices intra-act with pig bodies and microbes (Chapter 5), or the way local authority budgets intra-act with food safety inspections (Chapter 6), or disease publics intra-act with health advice (Chapter 7) or birds and viruses intra-act with people (Chapter 8), all affect the disease potential of the situation. These intra-actions may amplify or reduce disease risk, they may mask one problem while generating another, they may, in short, intra-act in various ways and with various effects.

In order to study disease, therefore, we need to both take and develop a geographical imagination that is attentive to the spatiality of situations (Allen et al., 1998), and use an STS-style thinking that is rehearsed in studying the ways in which different modes of ordering (Law, 1994) and human and non-human actors hold matters together, keep them apart and more broadly affect the disease situation.

So we need to be able not only to refocus attention on the socio-technical disease set-ups, but also understand these situations, which are infiltrated with other pressures, issues and materials. In this sense, we are not simply shifting the *location* of concern (from what disease ecologists call the hot spot of the forest edge to the hotspot of the farm), we are doing what anthropologists Brown and Kelly (2014) call *locational* research. We are interested in the multiple spaces, or spatialities, of disease, the meeting up and formatting of economic, technical, biological and political pressures that can amplify or indeed mitigate a disease emergency. In order to say some more about this situated, or geographical, approach we will now introduce a spatial critique of disease and through this discuss the concept of pathological lives.

From Pathogens to Pathogenicity, and Pathological Lives

Infectious disease understanding and disease control tend to focus on pathogenic (disease causing) organisms or viruses, their vectors and the wildlife reservoirs that may host them. Dealing with these through early warning, surveillance, pharmaceuticals, on-site hygiene or physical borders between domestic animal life and a less orderly (wild and microbiologically promiscuous) world have become significant components of what is called biosecurity (Chapter 2). Hard landscaping and barriers are coupled with maps that detail movement and incursion as a means to deliver least cost disease prevention (Chapters 3 and 4). In turn, relatively little disease-related policy attention has been paid to hosts or their (socio-technical) environments.

In contrast, and in this book, we adopt and develop accounts of disease that emphasise the entanglements of bodies, microbes and infrastructures, and thereby a relational understanding of disease. Our intent is to expand on the potential for a pathological understanding of life, and a more continuous, less dichotomous sense of health.

In this sense, we engage, albeit carefully and critically, with a counter narrative to the anti-microbial or anti-life tenor of disease response. In this alternative, pathogens, or the microbial world more generally, are very much a part and parcel of life wherever it may be. This is a world not so much threatened by the microbial outside but one where the manner in which lives are made, and the ways in which the inevitable entanglements between hosts, environments and microbes are handled, are key to any prospect for safe and indeed good life.

This diagramming (Chapter 2) of the spaces of disease and health in a globalised food and farming system is an essential task for re-conceptualising the disease emergency. In order to take this further, we are using the term *pathological lives* to foreground our interest in socio-biological entanglements. We take the term from the philosopher Michel Foucault, and from Eugene Thacker's engagement with disease (Thacker, 2009), so it is worth briefly revisiting its initial use before outlining how we wish the term to be read.

Michel Foucault used the term 'pathological life' to register a shift in biomedical understanding that occurred around the turn of the nineteenth century. In his account, disease started to shift from being an external threat *to* life to one that was part and parcel *of* life. Disease and life started to be understood as 'bound together' (Foucault, 1973: 153). As a result, 'the idea of a disease attacking life [needed to be] replaced by the much denser notion of pathological life' (1973: 153).

Understanding 'morbid phenomenon... on the basis of the same text of life, not as nosological essence' (1973: 153) had spatial consequences. Inside and outside no longer supplied the spatial coordinates of life and disease. The result was an unsettling of the conventional mapping of bodies as discrete objects with clear boundaries. The 'familiar geometry' of the anatomical atlas with its 'lines, volumes, surfaces, and routes' (Foucault, 1973: 3) started to be undermined by this denser notion of pathological process. The map of the body, with it regions and borders, and with its voyages of disease entities into the centre, started to lose its explanatory power. For example, speaking of much later challenges that accompanied the growing knowledge of viral process, Foucault asked:

> Has anyone ever drawn up the specific geometry of a virus diffusion in the thin layer of a segment of tissue? Is the law governing the spatialisation of these phenomena to be found in the Euclidean anatomy? (Fourcault, 1973: 3)

The answer is evidently no, and the alternative is rather close to what we will refer to as a topological approach to disease (Chapter 3). The point is that a different spatial imagination for disease, one no longer constrained by conventional geometry, became both possible and important.

The immediate consequences of thinking in ways that refuse strict boundaries between microbes and hosts, or categorical demarcations between pathogenic and commensal microorganisms, are that disease becomes a relational achievement. It is an achievement in which *pathogenicity*, or the tendency to produce disease, is made through the particular configurations of microbes, bodies, environments and so on (Farmer, 2004). Pathogenicity is a process, rather than a fixed object. It involves the significant intra-actions of microbial populations, hosts, immune responses, and the particular entanglements of animals, people, microbes, economics and politics.

These relational achievements that produce pathogenicities are what we are calling disease situations. Good life no longer becomes premised upon the absence of illness, or of microbes or pathogens, but is the subject of specific interplays of bodies, microbes, infrastructures and practices. The result is a shift in geographical imagination away from a topographical epidemiology of spread, of presence/absence and of disease barriers, to a topological epidemiology of bodily and molecular deformations, disease expression and more or less healthy lives.

From Governing Life to a Livelier Politics

A key analytical tool for understanding the approaches that are made to manage disease, to regulate human and non-human populations and to intervene in 'life' is biopolitics. The term has a long and chequered history (Lemke, 2011) but was notably adopted by Michel Foucault (1981) to mark a shift in emphasis, around the seventeenth and eighteenth centuries in Europe, in the government of human societies. In brief, biopolitics for Foucault signalled a broadening of the techniques and apparatus of government from the disciplining of individual bodies to the knowledgeable manipulation of a population, of statistics and, crudely, the use of precise mechanisms and processes associated with broad-scale changes as a means to affect the direction of those changes (Foucault, 1981: 137). It involved 'the coalescing of disciplinary codes, population surveillance mechanisms, and discourses concerned with the production and protection of (mostly human, mostly Western) "life"' (Nading, 2013: 66). The broad story is that nascent life sciences and statistics, many of them honed in plant and livestock management, entered or became 'intricate with' politics in the guise of a range of regularised processes that could be more or less effectively arranged 'so as to optimise a state of life' (Foucault, 2004: 246).

The commonly noted point is that the term biopolitics as a mode of governing applies largely to the management of human society, and seeks to render the material and non-human world as something to be manipulated. The latter becomes mere matter made fit for human ends. In this sense, disease management strategies and biosecurity, which involve the regulation and regularisation of non-human life, and of farming and food practices as a means to produce and protect (mostly human) life, are clearly biopolitical in terms of style and substance. They seem to suggest and are often written about in terms of the will to exert control upon and power *over* both human and non-human life (including non-human animals but also viruses, genes, businesses and so on).

And yet, biopolitics is rarely if ever that simple – for in seeking to govern 'a complex of men and things' (Foucault, 2007: 96; Lemke, 2015) a tension arises between encouraging more interactions with the worlds of others while trying to exert control over the resulting imbroglios of people, goods, services, ecologies and so on. In this sense, there is a well-known trade-off between attempts to encourage the proliferation and expansion of economic life while at the same time safeguarding its existence. Bluntly put, the flows and circulations of economic life can be both good *and* bad. Moreover, too much security or regulation of those circulations can affect not only the unwelcome elements of contact and movement, but also those life affirming aspects that require the contingencies of contact. Life as a result can start to suffer (Dillon, 2007; Dillon and Lobo-Guerrero, 2008).

This conundrum is at the heart of *Pathological Lives*, which asks how these powers over life and powers of life are handled, regularised and organised? Again, how are the inevitable entanglements within and between lives understood and regulated? Is biosecurity, for example, as it is implemented through laboratories that seek to order organisms and within a food sector that pursues disease freedom, an example of an over-blown and self-defeating approach to the regulation of life? Are there alternatives that can make life safe in ways that are not predicated on a control over life?

The problem may be that, in conventional formulations and as a mode of analysis as well as a form of governance, biopolitics and the power to regulate or balance life, tends to invoke a division between worthy and less worthy forms of life, one that is rooted, many argue, in a Heideggerian foundational distinction between proper and improper life (Agamben, 2002; Campbell, 2011; Wolfe, 2013). It is a distinction that results in the separation of 'truly' human life from sub-human and animal life, and sanctions what Wolfe (2013) calls the non-criminal putting to death of human, domestic and wild non-human animals.

In this vein, biopolitics offers us something of a resource as well as a potential curse or at least a warning. The warning has two elements. First, making life safe always risks the very thing it seeks to protect. This is, it seems, inherent in the tension between proliferation and security and requires a watch over the tendency to assume a (human) mastery or power over life. Second, there is a continuous possibility that any affirmation of the powers of life (for example in celebrating the interactions between hosts, microbes and environments) can rather quickly be turned against itself. In asserting a positive aspect of life we risk, it seems, the re-elevation of the proper over the improper. Security, as Foucault conceived it, was never meant to be a stable rubric, identifying good and evil. Rather it was and is a continuously shifting, even 'grasping' (Lentzos and Rose, 2009), logic. The point to hold on to for now is that pathological lives are intended to be read against the grain of any tendency to homogenise, de-skill and disinfect the socio-technical disease set up in the name of security. But they also need to be framed in ways that are not reducible to new norms or re-configured powers over life.

The resources for avoiding this disqualification of the many (most lives) in the name of the few (proper life) are not easy to discern, but we draw in the main on two bodies of work. First, and within the biopolitical tradition, we use the immunitarian thinking of Italian philosopher Roberto Esposito to open up a broader account of good life. Second, we adopt the cosmopolitical approach of Isabelle Stengers as a means to specify the ecologies and practices that can re-diagram disease as a situated and more than human matter.

Esposito demonstrates how community is bound together with immunity (Esposito, 2008, 2011). In this sense, immunity is not a matter of violent defence of the self against foreign attack, but instead is a matter of continuous communi-cation. Immunity is a shared space, a learnt property that is conveyed through contact rather than separation. Self and other, human and non-human, are not, in this version of being, tightly bordered but intra-act. They form a kind of borderlands (Hinchliffe et al., 2013) and in doing so unsettle any foundational separation between the proper and the improper.

While this is attractive in that it disrupts any foundational or *a priori* identification of proper life, the problem may well be that we can flip from a tri-umphant anthropological account of human exceptional-ism, to a world of few if any distinctions. Living with all manner of others does little to specify *how* exactly those lives are to be lived. So while we are taken by the challenge of a continual process of building immunity within a shared space, it strikes us that the more pragmatic and minoritarian philosophical tradition that Stengers called cosmopolitics can offer us more resources for thinking through the practical and spatial politics of this relational being.

Unlike biopolitics, cosmopolitics for Stengers is expressly meant to refuse a potential unity or accession to some version of proper life. Cosmopolitics in this usage is not equivalent to the Kantian notion of a good life (Stengers, 2005b). It is not about using norms to define the good, or optimise life. Rather it is about using the slippages and challenges to life as a means to question those norms and to open up a political space for counter-norms.

As Michael Schillmeier (2013) makes clear, there is distinction to be made here between a Kantian cosmopolitics of health and what he calls a cosmopolitics of illness. While the former involves proposing a proper life in order to constitute a system of rights, the latter can invite us to pose questions about norms rather than insisting on their re-constitution. Following Canguilhem (1991 (1966)), ill-ness for Schillmeier is not a deviation from the norm, or something that tends to expel sufferers from the polis. Rather, illness becomes a kind of messenger, the parasite a noisy interruption (Serres, 2007), and a time for re-configuring norms. In this vein, the emergency diseases with which we started become more than simply a challenge to established routines and practices – they urge us to question those norms and routines.

Cosmopolitics is, as Stengers observes, 'far more to do with a passing fright that scares self-assurance' (Stengers, 2005a: 996, see also Schillmeier, 2013: 35) than a self-assured re-constitution of proper life. These passing frights invite us to compose life differently, and in ways that take seriously the challenges to living that are posed by non-human beings and their human spokespeople (Stengers, 2010b). In other words, this is a form of learning that takes the obligations to, the hesitations before and relations with others of various kinds, as matters that are not simply vital for life but key for a re-constituted politics of living.

So our compulsion in this book is to both use the resource of biopolitical thinking to inform our analyses of pathological lives (or ask how have matters of life been governed), but also to question its foundations and limits (ask how have these lives been thought and normalised). Certainly, the intrication of life and politics reminds us that life and liveliness are far from innocent terms, and are made and unmade through various interrelations. And yet, we are also aware that biopolitics may be a poor ally in working through a current politics of life, in which the distinctions between human and non-human lives are even less apparent, not least because of the rise of zoonotic diseases. Moreover, once we start to surrender the human-centred focus of biopolitics, it may well be that we start to question the assumed ends of such a politics (the optimisation of a particular version of life).

In this sense, Isabelle Stengers's 'cosmopolitics' (Stengers, 2010a, 2011) seems to allow a more significant role for the microbes, economic margins and other non-humans that inhabit the situations we report. More important still, those others have more of a chance, we would argue, of forcing us to rethink our current situation. They are not subject to more optimisation but are key players in shifting the terms of the politics that we find ourselves in.

This shift in the terms of engagement returns us to the emergency with which we started. Without denying the urgency that is the disease situation, we would like to critically engage the infectious disease emergency, disease emergence and other terms. The emergency in waiting will not disappear as this book proceeds, but we hope to shift the question or focus from dangerous pathogens to dangerous situations and in so doing open up what Bonnie Honig (2009) calls a new kind of emergency politics.

We now turn to disease diagrams and later to situations as means to further specify our approach to pathological lives.

References

Agamben, G. 2002. *The Open: Man and Animal*. Stanford, CA: Stanford University Press.
Alexandratos, N. & Bruinsma, J. 2012. *World Agriculture towards 2030/50, 2012*, update. Food and Agricultural Organisation of the United Nations, Agriculture and Economics Division http://www.fao.irg/docrep/016/ap106e/ap106e.pdf
Allen, J., Massey, D. & Sarre, P. 1998. *Rethinking the Region*. London: Routledge.
Amoore, L. 2013. *The Politics of Possibility: Risk and Security Beyond Probability*. Durham and London: Duke University Press.
Anderson, B. 2010. Preemption, precaution, preparedness: Anticipatory action and future geographies. *Progress in Human Geography*, 34, 777–798.
Anderson, W. & Mackay, I.R. 2014. *Intolerant Bodies: A Short History of Autoimmunity*. Baltimore, MD: Johns Hopkins University Press.
AVMA 2008. *One Health: A New Professional Imperative*. Schaumburg, IL: American Veterinary Medical Association.
Barad, K. 2007. *Meeting the Universe Halfway: Quantum Physics and the Entanglement of Matter and Meaning*. Durham, NC: Duke University Press.
Braun, B. 2007. Biopolitics and the molecularization of life. *Cultural Geographies*, 14, 6–28.

Brown, H. & Kelly, A.H. 2014. Material proximities and hotspots: Towards an anthropology of viral haemorrhagic fevers. *Medical Anthropology Quarterly*, doi: 10.1111/maq.12092.

Campbell, T.C. 2011. *Improper Life: Technology and Biopolitics from Heidegger to Agamben*. Minneapolis: University of Minnesota Press.

Canguilhem, G. 1991 (1966). *The Normal and the Pathological*. New York: Zone Books.

Collier, S.J. & Lakoff, A. 2008b. The problem of securing health. *In*: Collier, S.J. & Lakoff, A. (eds), *Biosecurity Interventions: Global Health and Security in Question*. New York: Columbia University Press/SSRC.

Dillon, M. 2007. Governing terror: The state of emergency of biopolitical emergence. *International Political Sociology*, 1, 7–28.

Dillon, M. & Lobo-Guerrero, L. 2008. Biopolitics of security in the 21st century. *Review of International Studies*, 34, 265–292.

Donaldson, A. 2008. Biosecurity after the event: Risk politics and animal disease. *Environment and Planning A*, 40, 1552–1567.

Esposito, R. 2008. *Bios: Biopolitics and Philosophy*. Minneapolis: University of Minnesota Press.

Esposito, R. 2011. *Immunitas: The Protection and Negation of Life*. Cambridge: Polity Press.

Farmer, P. 1999. *Infections and inequalities: The Modern Plagues*. Berkeley, CA: University of California Press.

Farmer, P. 2004. *Pathologies of Power*. Berkeley, CA: University of California Press.

Food Standards Agency 2010b. *Meat Industry Guide: Guide to Food Hygiene and Other Regulations for the UK Meat Industry*. London//http://www.food.gov.uk/business-industry/meat/guidehygienemeat Accessed 30 October 2015.

Food Standards Agency 2013. *A Refreshed Strategy to Reduce Campylobacteriosis from Poultry*, September 2013 http://www.food.gov.uk/multimedia/pdfs/board/board-papers-2013/fsa-130904.pdf Accessed 17 August 2015.

Foucault, M. 1973. *The Birth of the Clinic: An Archaeology of Medical Perception*. New York: Vintage.

Foucault, M. 1981. *The History of Sexuality*, vol. 1: *An Introduction*. Harmondsworth, UK: Penguin.

Foucault, M. 2004. *Society must be Defended*. London: Penguin.

Foucault, M. 2007. *Security, Territory, Population: Lectures at the College de France 1977–78*. London: Palgrave Macmillan.

Garnett, T. & Godfray, C. 2012. *Sustainable Intensification in Agriculture. Navigating a Course through Competing Food System Priorities*. Oxford, UK: University of Oxford, Food Climate Research Network and the Oxford Martin Programme on the Future of Food.

Garret, L. 1994. *The Coming Plague: Newly Emerging Diseases in a World Out of Balance*. New York: Penguin.

Garrett, L. 2013. The Big One? Is China covering up another flu pandemic – or getting it right this time? *Foreign Policy*, 24 April.

Harris Ali, S. & Keil, R. (eds) 2008. *Networked Disease: Emerging Infections in the Global City*. Oxford: Wiley-Blackwell.

Hinchliffe, S., Allen, J., Lavau, S., Bingham, N. & Carter, S. 2013. Biosecurity and the topologies of infected life: From borderlines to borderlands. *Transactions of the Institute of British Geographers*, 38, 531–543.

Honig, B. 2009. *Emergency Politics: Paradox, Law, Democracy*. Princeton, NJ: Princeton University Press.

House of Commons Committee of Public Accounts 2013. *Access to clinical trial information and the stockpiling of Tamiflu*, House of Commons, London http://www.publications. parliament.uk/pa/cm201314/cmselect/cmpubacc/295/295.pdf

Janes, C., Corbett, K., Jones, J. & Trostle, J. 2012. Emerging infectious diseases: The role of the social sciences. *Lancet*, 380, 1884–1886.

King, N.B. 2002. Security, disease, commerce: Ideologies of postcolonial global health. *Social Studies of Science*, 32, 763–789.

Landecker, H. 2015. Antibiotic resistance and the biology of history. *Body and Society*, doi: 10.1177/1357034X14561341.

Law, J. 1994. *Organizing Modernity*. Oxford: Blackwell.

Leach, M. & Dry, S. 2010. Epidemic narratives. *In*: Leach, M. & Dry, S. (eds), *Epidemics: Science, Governance and Social Justice*. London: Earthscan.

Lemke, T. 2011. *Biopolitics: An Advanced Introduction*. New York: NYU Press.

Lemke, T. 2015. New materialisms: Foucault and the 'Government of Things'. *Theory, Culture and Society*, 32, 3–25.

Lentzos, F. & Rose, N. 2009. Governing insecurity: Contingency planning, protection, resilience. *Economy and Society*, 38, 230–254.

Mason, K. 2014. Risky (Agri-)business: Risk assessment, analysis and management as biopolitical strategies. *Sociologia Ruralis*, 54, 382–397.

Maye, D., Dibden, J., Higgins, V. & Potter, C. 2012. Governing biosecurity in a neoliberal world: Comparative perspectives from Australia and the United Kingdom. *Environment and Planning A*, 44, 150–168.

McCloskey, B., Osman, D., Zumla, A. & Heymann, D.L. 2014. Emerging infectious diseases and pandemic potential: Status quo and reducing risk of global spread. *The Lancet: Infectious Diseases*, 14, 1001–1010.

Methot, P.-O. & Fantini, B. 2014. Medicine and ecology: Historical and critical perspectives on the concept of 'emerging disease'. *International Archive of the History of Science*, 64, 213–230.

Mol, A. 2002. *The Body Multiple: Ontology in Medical Practice*. Durham, NC: Duke University Press.

Morse, S.S. (ed.) 1993. *Emerging Viruses*. Oxford: Oxford University Press.

Nading, A.M. 2013. Humans, animals and health: From ecology to entanglement. *Environment and Society: Advances in Research*, 4, 60–78.

Omran, A.R. 1971. The epidemiological transition: A theory of the epidemiology of population change. *The Milbank Memorial Fund Quarterly*, 49, 509–538.

Pew Commission 2008. *Putting meat on the table: Industrial farm animal production in America*. Baltimore, MD: Pew Charitable Trusts and Johns Hopkins Bloomberg School of Public Health.

Quamenn, D. 2012. *Spillover: Animal Infections and the Next Human Pandemic*. London: Vintage.

Schillmeier, M. 2013. *Eventful Bodies: The Cosmopolitics of Illness*. Farnham, UK: Ashgate.

Serres, M. 2007. *The Parasite*. Minneapolis: University of Minnesota Press.

Shaw, G.B. 1909. *The Doctor's Dilemma: Preface on the Doctors*. London: Penguin.

Shukin, N. 2009. *Animal Capital: Rendering Life in Biopolitical Times*. Minneapolis: University of Minnesota Press.

Sloterdijk, P. 2013. The immunological transformation: On the way to thin-walled 'socie-ities'. *In*: Cambell, T. & Sitze, A. *(eds)*, *Biopoltitics: A Reader*. Durham and London: Duke University Press.

Stengers, I. 2005b. Introductory notes an ecology of practices. *Cultural Studies Review*, 11, 183–196 http://epress.lib.uts.edu.au/journals/index.php/csrj/article/view/3459/3597

Stengers, I. 2010a. *Cosmopolitics 1*. Minneapolis: University of Minnesota Press.

Stengers, I. 2010b. Including non-humans in political theory: Opening Pandora's box? *In*: Braun, B. & Whatmore, S. (eds), *Political Matters: Technoscience, Democracy, and Public Life*. Minneapolis: University of Minnestoa Press.

Stengers, I. 2011. *Cosmopolitics II*. Minneapolis: University of Minnesota Press.

Taylor, L.H., Latham, S.M. & Woolhouse, M.E. 2001. Risk factors for human disease emergence. *Philosophical Transactions of the Royal Society B: Biological Sciences*, 356, 983–989.

Thacker, E. 2009. The shadows of atheology: Epidemics, power and life after Foucault. *Theory, Culture and Society*, 26, 134–152.

Thornton, P.K. 2010. Livestock production: Recent trends, future prospects. *Philosophical Transactions of the Royal Society B*, 365, 2853–2867.

Virginia Health Bulletin, December 1908, 1(6), 216.

Wallace, R.G. 2009. Breeding influenza: The political virology of offshore farming. *Antipode*, 41, 916–951.

Webster, R.G. & Walker, E.J. 2003. The world is teetering on the edge of a pandemic that could kill a large fraction of the human population. *American Scientist*, 91, 122.

Wellington, E.M.H., Boxall, A., Cross, P., Feil, E., Gaze, W.H., Hawkey, P., HJohnson-Rollings, A., Jones, D., Lee, N., Otten, W., Thomas, C. & Prysor Williams, A. 2013. The role of the natural environment in the emergence of antibiotic resistance in Gram-negative bacteria. *Lancet: Infectious Diseases*, 13, 155–165.

Wolfe, C. 2013. *Before the Law: Humans and Other Animals in a Biopolitical Frame*. Chicago and London: The University of Chicago Press.

Wolfe, N. 2011. *The Viral Storm*. London: Penguin.

Wolfe, N., Dunavan, C.P. & Diamond, J. 2007. Origins of major human infectious diseases. *Nature*, 447, 279–283.

World Health Organisation 2007. *The World Health Report 2007 – A Safer Future: Global Public Health Security in the 21st Century*. Geneva, Switzerland: World Health Organisation.

World Health Organisation 2015. Warning signals from the volatile world of influenza viruses. *Influenza* http://www.who.int/influenza/publications/warningsignals201502/en/ Geneva, Switzerland: World Health Organisation.

Chapter Two
Biosecurity and the Diagramming of Disease

Infectious and communicable diseases have been and continue to be understood in a variety of ways. 'Falling' ill or succumbing to disease has been variously understood as an act of fate, divine retribution, a matter of individual temperament, a result of environmental conditions, of contact with animals and insects, and more latterly caused by the incursion of microorganisms. In the eighteenth and nineteenth centuries, people were likely to speak of fevers *taking hold* of a person (Anderson and MacKay, 2014). Falling ill would be explained by a sufferer's constitution, a key cause being an unbalancing of humours. These physiological and 'individual' accounts of disease often co-existed with more environmental or external explanations. Miasmas or bad airs were implicated in the onset of certain diseases. By the late nineteenth century, this externalisation of causes started to gain more ground. Germ theory or the association of communicable diseases with microorganisms displaced (but did not necessarily replace) some earlier theories. The name influenza, for example, is derived from the Italian, evoking the experience of being influenced by the heavens. In French and old English the term for influenza is 'grippe', which has similar connotations, conveying the sense of being at the mercy of, or grasped by, disease. The disease is now more often than not associated with highly prolific and mutable viruses.

These different ideas relating to communicable disease inform or shape the ways in which those diseases are managed. If the upper social classes sought to move to 'Bel Air' when miasmas were the ascendant explanation of infectious disease, then the sanitisation of domestic and other spaces became the preventive

Pathological Lives: Disease, Space and Biopolitics, First Edition. Steve Hinchliffe, Nick Bingham, John Allen and Simon Carter.
© 2017 John Wiley & Sons, Ltd. Published 2017 by John Wiley & Sons, Ltd.

act of the modern or mid-late nineteenth century, germ theory influenced period. From the mid-Victorian architecture of glazed tiles in kitchens, food halls and toilets, to the culling of badgers in the British countryside as an attempt to rid farms of the bovine tuberculosis bacteria that they may excrete, germ theory, or the association of infectious disease with microbial agents, has contributed to a hard landscape, to smooth surfaces and impermeable barriers.

In the current moment, germs or the molecular chemistry that describe virulence factors, remain prominent in accounts of infectious disease. Even so, we should not give the impression that there is or has been at any one time a single, totally dominant account of infectious disease. Indeed, social medicine, studies of immunity and autoimmunity, and clinical practice often rely on explanations for disease that draw us back to biographical details, physiology and ecology (Anderson and MacKay, 2014; Methot and Alizon, 2014). Accounts of communicable disease in this sense seem to involve a continuous ebb and flow of arguments that shift the locus of explanation from the body, to outside, from environments to molecular structures, and from social conditions to the entry, from without, of rogue molecules.

To conceptualise these differences and shifts, we will use the term 'diagrams of disease'. Our overall aim is to use this phrasing, with its explicit spatialisation of disease understandings and practices, to develop an account of how the term biosecurity enacts (or engages and performs) disease in particular ways. While inherently spatial, diagrams are not cartographies or mappings of disease. They are instead bundles of concepts and practices that inform understandings of communicable disease and interventions in disease management. So diagrams help us, first, to establish key understandings of and techniques that relate to disease – how diseases have been matters for different kinds of intervention and spatial strategy (like exclusion of sufferers or inclusion into public health). Second, in practice there is likely to be more than one diagram in play. Here it is the tension within and between diagrams as they interfere with one another or are coordinated that becomes the key issue. How these tensions are resolved or coordinated helps to generate what we are calling disease situations, a focus for Chapter 3.

We start the chapter by outlining how infectious disease can be, variously, something that requires a *division* of society (into healthy and ill); something that requires social *organisation;* and something which requires *intervention*. In the first section below (Disease Diagrams), these three disease diagrams allow us to delimit the various influential ways in which diseases have been and are understood and addressed. The second section (The Disease Multiple: Germs and the Return of the Outside) supplements this account of disease diagrams with a more empirical, science and technology studies (STS) derived approach, wherein disease becomes a multi-diagrammed issue. Here we use the same three diagrams to assess the rise of microbes in causal explanations of infectious diseases and, in doing so, note how disease is managed in ways that draw on more than one diagram. Disease is, in this sense, a multiple matter (Mol, 2002). Finally, in the last section (Biosecurity and the Diagramming of Disease), we explore what such an understanding of disease as multiple implies for current efforts to reduce the risk of or prepare for infectious disease events and emergencies captured by the mixed bag of policies and measures referred to as 'biosecurity'. How, in other words, does this mix of diagrams play out in the current climate? To do this, we work through five key elements that help to characterise biosecurity, and which – as we shall demonstrate – pull and stretch disease in different (and sometimes competing) directions. The result is that infectious disease as it is addressed by biosecurity exists as matter in tension. How these diagrammatic tensions are dealt with, lived with or otherwise handled becomes a key component of what we term, in Chapter 3, disease situations.

In teasing out different understandings of and approaches to infectious diseases, we will emphasise the overlaps and interferences within and between them. It is not so much the hard and fast distinctions between viewpoints, ideologies or particular drivers of activity that interest us here. Instead, we are interested in their co-existence and the effects this has on infectious disease practices. Our aim for this chapter is to generate greater understanding of the tensions that can exist between disease practices. Only by identifying these tensions can we hope to understand their role in producing disease situations and look, ultimately, to build better pathological lives.

Disease Diagrams

To use a diagram is often to picture something, to lay out components on a sheet of paper and to envisage key components and their relationships. In social theory, it can also evoke the ways in which matters and issues 'come to light', and are spoken and practised, or enacted.

Diagrams, in this sense, are the discursive and non-discursive 'mappings' into thought and practice that give shape to an issue or problem. Privileging neither ideas nor actions, they inform how an issue is brought to attention, understood and enacted (Deleuze, 1999; Foucault, 1977). A diagram is not a representation, but it is a style of bringing the world into being, a bringing forth that, it should go without saying, involves people, architectures, non-human actors and so on. Diagrams are never perfect, or complete. They are best thought of as ongoing processes that combine many elements together. Moreover, as we will see, there may be more than one diagram that relates to a particular set of issues. In that sense, the issue or object will not only be understood in different ways but also practised or configured through varying combinations of diagrams. It is possible, with more than one diagram, for people to say and do different things, for diagrams to compete with one another, or, just as likely, for diagrams to co-exist and work together. In this sense, the ways in which diagrams interrelate or are coordinated becomes a key topic for study and analysis. It allows us to understand how and why certain courses of action are followed and how we may develop some leverage on those actions.

Importantly, a diagram is always a work in progress, a 'writing around' (Hinchliffe, 2007) a topic, and therefore informs but does not exhaust that topic or the ways it can be imagined and enacted. The diagram is, as Deleuze had it, 'highly unstable and fluid, continually churning up matter and functions in a way likely to create change' (Deleuze, 1999: 35). Diagrams in this sense are not of history (or related to a particular epoch) but help to *make* those histories by 'unmaking preceding realities' (ibid.). They are, for Deleuze at least, abstract machines, driving continual evolution. Diagrams may be the conditions for but not the be all and end all of action. In these senses, disease diagrams can help us to sketch out and then assess the common ways in which infectious diseases are conceptualised, framed and addressed, but also identify room for manoeuvre and possible change.

Foucault used three 'paradigmatic' diseases and the characteristic political and/or religious technologies with which they are associated as model diagrams. They are leprosy in biblical accounts, plague in medieval and early modern Europe and smallpox in the nineteenth century (Foucault, 1977, 2004). As is the case in much of Foucault's work, the diagrams are drawn from historical cases. However, as we noted already, they need not be read as defining historical config- urations but instead can mark abstracted modes of being or political technologies through which key concepts and ideas are both thought and imperfectly prac- tised. The same can be said regarding the modes of political organisation and power that Foucault relates to his paradigmatic diseases, and which were central to his broader project. In brief, these are sovereign or juridical power (where legal codes are enacted and enforced from the 'head' of state); disciplinary power (where institutions help to internalise norms); and biopolitics (or an apparatus of security) (Thacker, 2009). As Thacker notes:

> The first operates through interdiction and punishment, and thus law becomes central for sovereignty (which, for Foucault, is directly related to the capacity to punish or kill); the second operates not according to law but according to observation, surveillance and correction; and the third operates not through law or correction, but through means of calculation and intervention (Thacker, 2009: 140).

In *Discipline and Punish* (Foucault, 1977) and then in his lectures at the College de France (2004, 2007, 2008), Foucault uses infectious disease and epidemics to trace out these modes of power. The three paradigmatic cases illustrate the accre- tion (rather than a progressive replacement) of these modes and their accompa- nying disease knowledges.

Foucault (1977) initially compares the treatment of lepers in biblical times with the disease practices that were characteristic of medieval plague. Sufferers from leprosy, a communicable disease with clear and visible disease signs, were famously excommunicated or forced into exile, and cast out into the wilderness and a living death. In contrast, the plagues of early modern Europe were characterised by something other than this expulsion and division of diseased from healthy bodies. Certainly there *are* accounts of exclusion, with 'plague ships' refused landing rights (Artaud, 1970), and folkloric tales of pipers and rat removals, but plague was also characteristically a disease of organisation that involved enclosure and inclusion rather than simply exclusion.

Using an account published at the end of the seventeenth century of the order that would prevail should plague appear in a town, Foucault noted how the ensuing lockdown would freeze space, allow regulation to penetrate the smallest details of everyday life, and would forbid people leaving home or going about daily business – it was a system of permanent registration, transmission of reports and centralisation. The only things to circulate were documents bearing 'the name, age, sex of everyone, notwithstanding his condition' (Foucault, 1977: 196). Compared to the dream of purity that followed the exile of the leper, this was, for Foucault, the dream of surveillance. 'The plague stricken town [is] traversed throughout with hierarchy, surveillance, observation, writing...' (Foucault, 1977: 198). While the human carrier of disease had clearly been and might remain a life that could be sacrificed in order to restore good health, the dream of surveillance emphasised another dimension to dealing with disease. In short, the political dream of the plague, as Foucault called it, involved not so much 'laws transgressed, but the penetration of regulation into everyday life (...) [and] the assignment to each individual of his "true" name, his "true" place, his "true" body", his "true" disease' (Foucault, 1977: 198).

Thacker draws out the distinctions clearly:

> With leprosy, the political-theological response is to exclude and divide, and this is accomplished through rituals by which the leper is sent outside the city to a colony and pronounced by the Church as 'dead among the living'. Plague provides a slightly different case. If the political response to leprosy is to exclude and divide, then with plague the response is instead to include and organise. For example, in the mid-14th century, during the Black Death, many Italian city-states set up temporary public health committees to establish quarantine of towns and ships at port, as well as to account for the ill and the dead (Thacker, 2009: 141).

Quarantine, or the temporary holding or lock down of a person, shipment, group or non-human animal, clearly retains the diagrammatic echo of expulsion, but it is accompanied by registration, surveillance and the potential re-integration of the affected and infected bodies once the dangers of contagion have passed. This is not only a matter of sovereignty, or the sanctioning of living death, but also the disciplining of bodies through the emerging institutions of health and sanitation and their spatial and temporal organisation.

In the third disease diagram, concerning smallpox, another set of operations comes to the fore. If excommunicating lepers involved a sovereign right to disqualify life (and sanction living death), by casting out sufferers, and plague ushered in a technology of surveillance and ordering, then the development of a treatment for smallpox with the cow pox bacterium, or the first vaccine (from the Latin *vaccinus*, pertaining to a cow), marked another moment in the development of disease management. Inoculation involved a new kind of anticipatory approach to disease, one that mimicked the disease while reducing its effects. Vaccination and the development of meeting a risk with a risk, and the concomitant generation of a risk pool or population as a target of public health, started to offer new modes of anticipatory governance of disease.

For Foucault, this heralded a biopolitical set of operations where the focus is on intervention and calculation. Life is not so much divided into good and bad but is rather marked by continuous and risky movements, interactions and circulations – and these circulations are both what makes life possible as well as acting as the conduits for its endangerment (a point to which we return in the final section, Biosecurity and the Diagramming of Disease). Table 2.1 summarises the main elements of these three disease diagrams.

The three diagrams (exclusion, inclusion and normalisation) provide a first pass at the range of techniques and powers that relate to disease. In this historical telling, they suggest something of an evolution. However, it is important not to lose sight of the interactions between diagrams and their similarities as well as differences. Our task, then, is not to mistake a list of diagrams with a neat progression or history, but to interrogate their interrelations and co-existence. In short, the table does not quite capture the sense that diagrams co-exist, that they evolve and may well interfere, positively and negatively, with one another.

Here it is useful to draw on the work of scholars in a broad field called Science and Technology Studies. It is a body of work that amongst other things supplements Foucault's accounts of the practices and materialities of power (Law, 1991). It does so through detailed empirical investigations that tend to demonstrate how, *in practice*, matters are always heterogeneous. That is, rather than one mode of ordering or diagram achieving dominance, there tend to be numerous orderings and materials that together produce the social world. It is the

Table 2.1   Foucault's three diagrams.

	Leprosy	Plague	Smallpox
Diagram	Exclusion	Inclusion	Normalisation
Action	Divides	Organises	Intervenes
Technique	Banishment	Quarantine	Vaccination
Ontology	Religion/law	Political economy	Public health
Power	Law/sovereignty	Discipline	Security

(Thacker, 2009: 142. Reproduced with permission of Sage Publications.)

combination of orderings and materialities, or diagrams in our sense, that produce powerful effects.

To understand how more than one disease diagram can co-exist, we will consider the rise to prominence of microbes and pathogens in disease explanation, and in particular the rise of Louis Pasteur's interventions in microbial life. As we will see, the invention of pathogens in the nineteenth century helps us to demonstrate how *normalisation* and *inclusion* remain dependent on *exclusion*, and that an 'outside' continues to haunt the resulting disease multi-diagram.

The Disease Multiple: Germs and the Return of the Outside

For the anthropologist, Heather Paxson, people in the US (and one could add in the Global North more generally) now live in a Pasteurian world. It is a world where microbes are considered the prime causes of disease. Many, she suggests, 'blame colds on germs, demand antibiotics from doctors, and drink ultra-pasteurised milk and juice, while politicians on the campaign trail slather on hand sanitiser' (Paxson, 2008: 15). We will have cause to trouble this characterisation of public understandings of disease in Chapter 7, but the broader point holds: Pasteur's laboratory not only managed to make microbes visible, it also helped to change the world.

Pasteur, the nineteenth-century French chemist, is widely credited of course with the scientific and practical demonstration of germ theory, the idea that infectious diseases and the spoiling of produce are caused by microbiological organisms. In so doing he refuted the then prevalent notion that spoiling occurred from within, *sui generis,* or indeed was solely a chemical reaction in accordance with Liebig's anti-vitalist chemistry (Latour, 1999). Germs were to become the first causes or agents of the material and biological changes associated with spoiling and with communicable diseases.

As Latour's (1988) detailed, archival, engagement attests, Pasteur's achievement not only required innovations in the laboratory, it also involved shifts in the ways in which social and economic life was organised. To pursue the gains of germ theory, he required a network of hygienists, farmers, equipment, domestic animals and so on to trial and demonstrate techniques for controlling microbes in the field. In other words, germ theory was not simply a scientific idea that described how things really were and simply diffused out from Pasteur's laboratory. Rather, it was a theory that was built up through the laborious alignment and association of numerous human and non-human actors. The achievement of these associations was to make possible a world of social *and* microbial control, or the rational ordering and prediction of microbiological life. Making microbes visible thus marked, for Latour, a shift in authority and a rebalancing of powers:

> The asymmetry in the scale of several phenomena is modified: a microorganism can kill vastly larger cattle, one small laboratory can learn more about pure anthrax

cultures than anyone before; the invisible microorganism is made visible; the until now uninteresting scientist in his [sic] lab can talk with more authority about the anthrax bacillus than veterinarians ever have before (Latour, 1983: 146).

As Latour continues:

The change in scale makes possible a reversal of the actors' strengths; 'outside' animals, farmers and veterinarians were weaker than the invisible anthrax bacillus; inside Pasteur's lab, man [sic] becomes stronger than the bacillus and as a corollary, the scientist in his [sic] lab gets the edge over the local, devoted, experienced veterinarian (Latour, 1983: 147).

This power of making visible, and the levering up or raising of the world, as Latour calls it, is a trial of strength that also has profound effects on disease practices. It is a 'system of light' that ushers in a different diagram of disease, one that is clearly about inclusion (discipline) and normalisation (security) (see Table 2.1). But, it is also one that retains some familiar spatial coordinates. In particular, a will to divide the world (into pasteurised and unpasteurised) was *retained and even bolstered* as microbes were diagrammed as outsiders to healthy lives.

Indeed, the extent to which Pasteur's germ theory of contagion became accepted over the course of the nineteenth and twentieth centuries owed much to a desire to 'see' and drive out the agent of ill health (Canguilhem, 1991 (1966)). In other words, once microbes were isolated and treated as prime suspects in disease generation (or pathogens), the focus of attention shifted from the patient or sick organism to the threats that existed outside of the body. As the extracts from Latour's account also suggest, there is a shift in expertise from a hands-on, 'local' and 'experienced' veterinarian or farmer, attendant to the bodies and signs of disease, to the laboratory scientist interested in principles, microbes and the procedures needed to reduce their effectiveness in generating disease (see Chapter 5 for a rather different ecology of practice).

Our point for now is that pasteurisation entails more than one diagram. There is the discipline or the laboratorisation of food production, something that Latour's actor networks emphasise. There are calculations and interventions in the form of inoculations against anthrax. But germ theory also rekindles a sovereign and legalistic notion of disease. Indeed, the driving out of microbes became more than something that farmers and food producers signed up to as a matter of self-interest. Some infectious diseases and their microbial agents became notifiable and matters of legal jurisdiction once their presence had been confirmed. They became matters to be excluded.

The continuing valency of sovereignty or interdiction, even as the powers of 'observation, surveillance and correction' (more often associated with an inclusive diagram) were in effect, rested on a particular understanding of microbes. The latter drew on Koch's postulates – a set of experimental guidelines for identifying pathogens formalised by Robert Koch, one of Pasteur's contemporaries (Gradmann, 2009). In brief, the postulates held that a disease had an aetiology or causal pathway associated with a particular microbe if:

- the microbe could be shown to be present in abundance in diseased organisms (and absent in healthy ones, notwithstanding asymptomatic carriers);
- the same microbe could be isolated from the effected organism and cultured;
- the isolated microbe could then cause disease when introduced into a new organism; and
- it could, in turn, be isolated again and shown to be identical to the first generation of microbes.

While the postulates can hardly be said to be dogma in contemporary biology, they nevertheless help to diagram infectious disease in particular ways. They suggest that self-similar microbial organisms are responsible for causing disease to otherwise healthy organisms. They assume a dichotomy between a previously healthy inside and a pathogenic outside, with the crossing of the border between the two as a key moment of infection and disease. The resulting division of healthy bodies and disease bearing microbes presupposes a world of discrete and definable entities, with intact surfaces that may or may not come into contact. Prior to contact, there are coherent bodies with defined volumes.

This is a world of pre-existing physical objects that collide and have effects. It is a world, in short, with a particular understanding of dynamics and space that is made sense of in accordance with Euclidean geometry. These objects (pathogens, healthy bodies, infected bodies) move *over* a smooth space and once in contact cause infection and disease. As with most versions of space, a particular version of time is also reproduced:

> The requirements of the [micro]organism's isolatability and stability presume that what constitutes the organism must be already established and independent of the capability to infect the host. The requirement for 'pure culture' assumes that infection happens under all circumstances independent of any environmental variations (Schrader, 2010: 291).

In other words, this is an essentialist and ahistorical view of microbes and pathogens (Methot and Alizon, 2015), one governed by the rules of species type. The latter is defined through functional characteristics that are discerned in the laboratory. Pathogenic microbes are, to all intents and purposes, self-similar objects that reproduce along species lines following laws of hereditary wherein character and function are maintained. They are not, it should be noted, relational beings whose form and function are considered to be highly variable depending on the settings within which they reproduce (Chapter 3). They are, instead, like bullets, packages of disease that will cause infection and ill-health.

This fixed space-time of the microbe provides the basis for anti-microbial warfare, and the driving out of the agents of ill health. The division between healthy bodies and pathogenic microbes sets up a normative agenda whereby a dangerous and disorderly pathogenic world is distinguished from a virtuous, germ-free, world. This virtuosity involves, as we have said, a mix of disease diagrams, but it is evidently based on a particular geometrical or spatial-temporal imagination. The bright lines of battle are clearly drawn and need to be physically reinforced.

To be sure, and in practice, and as we have suggested, germ theory was always more than a matter of only excluding these self-same pathogens. Ludwig Fleck's (1979) detailed engagement with changing explanations for syphilis and the sero-logical diagnosis of disease in the 1930s demonstrated the arts of scientific work. Fleck's account clearly shows that most diagnosticians 'understood the warfare metaphor for illness was not how biology worked and cannot explain why some infected people fall ill and others do not' (Fischer, 2012: 144). Indeed, diagnosis or 'knowing around' disease required diagramming (or writing around). Our point though is that germ theory did not involve a progression of intervention at the expense of exclusion – understanding and acting on disease in the era of germs involved more than one diagram. The outside remains even if it is now in tension with other ways of knowing and doing disease.

It is the lasting legacy of germ theory that it retained and rekindled a diagram of disease as something that is distinct from healthy bodies, something to drive out. It is a legacy that is troubled by experience as well as scientific understand-ings of disease, but one that nevertheless continues to inform contemporary prac-tices. It is to these practices that we now turn.

We have so far traced three disease diagrams and used accounts of germ theory to suggest that disciplines of hygiene and calculative interventions in disease systems co-existed with the interdictions and exclusions that were associated with the identification and characterisation of pathogens as self-similar agents of disease. In other words, we have established that there is a mix of diagrams in play.

Our question, given this mix of diagrams, is to ask how these diagrams com-bine and shape disease practices. In the next section, we turn to contemporary biosecurity practices and identify five key elements that are associated with the term. Our aim is to present these elements in order to ask what work they do in re-diagramming disease. Our point, to repeat, is to identify the tensions that exist in disease practices as a means to understand and imagine them otherwise.

Biosecurity and the Diagramming of Disease

As we noted in Chapter 1, biosecurity is a term that is often used in relation to the threat of emerging infectious diseases and the raft of measures that govern-ments, commercial organisations, laboratories and others put in place in order to reduce disease risk and/or prepare for any consequences of a disease outbreak. In order to understand how infectious and food-borne diseases are managed, we need to unpack biosecurity and in doing so identify the ways in which various disease diagrams feature.

A key question is, given our excursions using Foucault and Latour, how do the various elements of biosecurity serve to re-orient infectious disease by drawing on and possibly altering various disease diagrams? We will answer this question by first saying a little more about biosecurity and its geographical as well institutional variability, before identifying some of the key ways in which biosecurity is framed through five elements. Detailing the latter allows us to draw out the continuing diagrammatic and spatial tensions in biosecurity practices.

Biosecurity has undoubtedly been framed by a series of events and coloured by its application within various domains. Terrorism attacks on the US in 2001, including the unrelated but associated anthrax attacks; the pre-emptive strikes on Iraq and Afghanistan by the US and its allies that followed; the 'paradigmatic'. Severe Acute Respiratory Syndrome (SARS) events of 2002, wherein global health systems seemingly contained a possible pandemic; the epizootic outbreak of foot and mouth disease in the UK in 2001, as well as the development of endemic bovine tuberculosis in cattle; the virus gain-in-function experiments reported in 2012 that caused concern not only about the highly contagious avian influenza viruses produced, but also the release of the biological information or genome into the public realm; the Ebola Viral Haemorrhagic Fever outbreak in 2014 in West Africa with its plague-like lock down.

With these and other events in mind, Collier and Lakoff (2008b) highlighted four broad domains in which biosecurity is used as a term: emerging infectious disease; bioterrorism; the cutting edge of life sciences; and food safety. Each is largely focused on microbial threats in a mobile world of rapid environmental and socio-political change, with the deliberate manipulation and/or dispersal of microbes, the inadvertent release of laboratory-based organisms (or their genetic 'information') and the vulnerabilities of a mass food production system as core concerns. Fears provoked include epidemics and pandemics, epizootics, crop diseases, food-borne illnesses, threats to vital systems and the 'perfect storms' that are generated through inter-system failures that may include promiscuous and exuberant microbial life as one of a number of triggers. A fifth domain in which biosecurity has been invoked would include the more macro-biotic realm of invasive plants and animals, a domain that together with microbial diseases of trees (Tomlinson and Potter, 2010), draws in broader fears over landscape and ecological security (Barker, 2010).

In the past two decades, a crude geography of biosecurity may well approximate to the following: a concern with global public health, terrorism and laboratory science dominates in the US, while Europeans have been particularly concerned with food sector vulnerability to disease, and Australasians have been most active in the area of preventing the incursion of invasive species (Hinchliffe and Bingham, 2008). Within Asia and Africa, concern has often focused on the threats to livelihoods posed by the diseases themselves, as well as what some might see as the disproportionate responses of Western-based interests in reducing the emergence of those diseases at the expense of mundane but more significant health concerns (Braun, 2007; King, 2002; Scoones, 2010; Leach and Dry, 2010). Just as there are broad geographical differences in emphasis and concern, there are also important sectoral variations, with national and international public health bodies like the World Health Organisation (WHO) advocating pandemic preparedness and capacity building (World Health Organisation, 2007), while veterinary and animal health organisations (in the guise of the Office International des Epizooties (OIE) or the World Organisation for Animal Health) advance disease prevention and eradication of threats (often as a means to conform to international trading rules). International development and food organisations, like the Food and Agricultural Organisation of the United Nations (FAO), meanwhile, have tended to juggle the concern to modernise or improve production systems with the need to safeguard livelihoods and sustenance (Chien, 2012; Scoones and Forster, 2011; Keck, 2008).

Biosecurity is clearly a composite term that has been applied in different settings. It works, as we will see, by drawing on and re-drawing a number of disease diagrams. In order to identify these and to think through the ways in which they have been variously combined, we now work through five key contemporary components or elements of biosecurity. First, there is the generic issue of securitisation; second, liberalism; third, an ontological shift to biological mutability; fourth, a pathogen-based logic to disease prevention; and, fifth, an anticipatory logic to infectious disease management. Many of these components make use of and accentuate different diagrams of disease and, as such, they can exist in tension with one another. As we show, the composite character of biosecurity ends up being rife with spatial tensions. How these are dealt with becomes a key component of what we are calling disease situations (Chapter 3).

Security

Biosecurity would be expected to share something with broader notions of security, and in the terms that we have used here, with Foucault's biopolitical or normalisation diagram of power. We will take each in turn.

First, security, in the broad sense, may usefully be distinguished from defence. If the latter is about stopping a known enemy, security often indicates an altogether broader set of activities and states of feeling or affect. Security, then, is not simply a matter of containment or shoring up existing (nation-state) boundaries. Rather, security is more of a process, and may involve building new borders within and beyond the nation-state, acting to prevent and pre-empt danger, and shaping the environment through constant activity (Hardt and Negri, 2005: 20).

Security, as an affective state, is constantly under construction and in that sense suggests a relentless mode of being.

This dynamic of continuous vigilance, shaping and developing assurance and re-assurance, also relates to a mode of power that is less concerned with interdiction or correction and more interested in calculation, intervention and an approach to threats that exist outside the normal predictive calculus of probabilistic risk. Lentzos and Rose provide a useful characterisation that emphasises these shifts in focus:

> Security (…) does not classify phenomena according to a fixed grid of good and evil, does not try to control and eliminate all infractions, but regards variations as inescapable in natural phenomena. This logic does not operate according to the binary of permitted and forbidden, does not judge a variation as evil in itself, but tries to grasp the reality of the 'natural' phenomena that it addresses, to understand the way in which various components function together, to manage or regulate that complex reality towards desired ends (Lentzos and Rose, 2009: 232).

In security terms, and somewhat departing from aspects of the basic germ theory model, a microbe is neither good nor evil, permitted or forbidden – it is rather part of a set of relations that must be engaged if the desired end of good health, in this case, is to be secured. Inoculation and vaccination, where the poison becomes the cure, provide the exemplary cases for this approach to securing life. But this enhancement of the ability of a body or a state to define self and non-self is a mixed affair. Contemporary security involves investment in:

> …border controls, regimes of surveillance and monitoring, novel forms of individuation and identification, notably those based on biometrics, preventive detention or exclusion of those thought to pose significant risks, massive investment in the security apparatus and much more (Lentzos and Rose, 2009: 231).

While Lentzos and Rose rightly question any uniform narrative and telos, 'security' clearly acts as a claim on resources within an expanding space of economic activity and circulation. This latter correlation between security and expansion is key. The more an economy or society expands, the more it participates in and helps to configure the circulation of goods, materials, ideas and so on. And the more these matters circulate, the more calls there are for securing or modulating those circulations. The point for now is that security implies a dynamic of constant activity and is an expanding domain within particular forms of economic activity.

Second, Foucault notably identified the rise of security with a biopolitical mode of bio-power (Foucault, 1981). The latter, in the main, refers to the governance of a risk pool or the statistical mapping and subsequent regulation of the individually random though, once aggregated, predictable attributes of a population. This installation of 'security elements around the random element inherent in a population of living beings served the desired end of *optimising* a state of life' (Foucault, 2004: 246, emphasis added).The patterning of the aggregated character of a population (notably its mortality, morbidity and fertility rates) was key in allowing nascent states to intervene biopolitically in how life was to be lived (Foucault, 1981).

An important point here is that a population, or risk pool, is not equivalent to a people. Rather, a population is made up of and is continually affected by the expanding space of circulations within which it exists (Foucault, 2007; Dillon and Lobo-Guerrero, 2008) and its regulation both becomes possible and is continually challenged and enabled by the very relations and circulations that make the population 'thrive' (Foucault, 2004). Meanwhile, as population becomes the key site for political intervention, circulation becomes the paradigmatic space for biopolitics, and its regulation a major pre-occupation. In short, the issue of how to manage circulations becomes a key concern for the government of society.

Security is, in this broadly Foucauldian account, part and parcel of the liberal extension and intensification of forms of exchange and trade. Expanding economic activity opens up new matters for securitisation, and its continuance depends on technologies of sorting, categorising and importantly encouraging some things and people to circulate while detaining or modifying others. From plague stones, used in the sixteenth and seventeenth centuries to allow coins to circulate and trade to continue in times of plague, to the boot dips that sprang up all over the British landscape following the foot and mouth outbreak of 2001, to immigration controls, airport and web-based security, there is a common theme of both facilitating and intervening in circulations. Security, in this sense and in the words of Yates-Doerr (2015: 112), concerns 'those infrastructures through which travel is orchestrated', a phrasing that captures this dual interest in movement and assurance. It also suggests why the relentlessness of security is so pervasive in a world that seems to be continuously on the move.

The exact interplay of these interventions, the borders, surveillance, detentions and regulations, Lentzos and Rose suggest, are reconciled in a variety of ways. Indeed, as we have noted, there is geography to this in terms of state proclivities and tendencies, but in the last instance (for Lentzos and Rose) governing is shot through with 'the two fundamental imperatives for those who would govern a liberal society today – the imperative of freedom and the imperative of security' (Lentzos and Rose, 2009: 232). These imperatives draw us to our second element, liberalism.

Liberalism

The logic of orchestrated movement and the dual imperative of freedom and security are underpinned by what many authors characterise as liberal or neoliberal approaches to governance.The latter often refer to the reorganisation of state and private sectors, with a greater role for non-state actors and market-oriented practices that exist either separately to or in new kinds of partnerships with state organisations. The links between liberalism and security are forged, the argument goes, through expansion (see above), changes to the organisation and management of, or responsibilities for, circulations, and the rise of *homo oeconomicus*, or a subject of interests. We will treat the latter two in turn.

As the complexities and intensities of circulations become ever more manifest in increasingly globalised societies, the already mentioned formatting of city/state activity through regulation of circulation tends to generate a financial and organisational need to re-distribute the work of regulation (and the risks and costs of bad circulations) more widely. Circulations become not just the concern of the state but involve and require action from a wide range of actors, including civil society. Coupled with a set of commitments to shrinking the state and re-distributing public finances, there is, in this account, a *centrifugal* aspect to security, with an outward shift in actions and responsibilities.

In the security arena, this extension of powers is often interpreted as a privatisation of both the state armature and welfare (Cooper, 2008) and a move to a re-distributed responsibility for risk reduction and cost sharing. So, for example, the desire to promote safe circulations coupled with a will to re-direct state expenditure on regulation towards private firms, and to reduce the red tape associated with regulating business that in this form of logic should be free to invent and expand, drives a shift towards greater upstream responsibility for the prevention of disease events and transmission. The injunction to modernise food production and agricultural practice is, in this sense, a sharing and modulating of responsibilities for public health. The rise of new public health, with shifts from state provision of health services to an individualised responsibility for one's own and dependents' health, is another exemplar.

This redistribution of agencies has both positive and negative consequences, as we will see in later case studies. But it would be wrong to suggest that the liberalism/security coupling is entirely captured by a centrifugal tendency. Indeed, the logics of impending if uncertain threat and wide-scale vigilance are also, it should be noted, backed up with the installation of legal and other measures to ensure and protect a capacity to re-centre command and control in emergency situations (Anderson and Adey, 2011; Collier and Lakoff, 2008a, 2008b). In addition, there is also a more twisted set of relations between what might formerly have been more readily identified as state and non-state actors. For responsibilisation is coupled to a new kind of political subject, one that is no longer only juridical (the subject of rights) but is now also a subject of interests (Wolfe, 2013). This new subject (*homo oeconomicus*) required a different kind of association to the state – one that is at arm's length and co-operative rather than bluntly coercive. Bio-power, with its focus on discipline and regulation, emerges in the effort to govern for, or on behalf of, the desired ends of those liberal subjects and their interests. It is interesting to note that those interests are now part and parcel of governance and security, and the result can be not only a centrifugal distribution of responsibilities but also a dynamic re-distribution of the costs and benefits of security. The privatisation of security, in the form of health care services, policing, border agencies, safety inspectorates and so on, introduces a logic that can result in benefits to shareholders at the same time as increasing the exposure of certain (often non-marketable activities and people) to risk and threat. In other words, while some people seem to benefit from the ethopolitical world of being able to make healthy choices and pursue an optimised life, many others (people and non-human animals) may, as a result, be relegated to serving those privileged lives. The subject of interests is conditional upon the objectification of others who may be relegated to precarious and bare forms of life (Braun, 2007).

With this contradictory liberal landscape, disease is pulled in a number of directions. Acute events may well remain a matter for sovereignty and interdiction, but there is also a move to shift the burden of responsibility towards non-state actors. At the same time, the colonisation of governance by a subject of interests and the marketisation of security can result in a more uneven distribution of disease burden and costs.

Mutable Worlds and Emergent Microbes

In recent years, this liberal tradition, and the accompanying rise of non-state or quasi-state actors, has been aided and abetted by the shifting knowledge claims regarding existing and anticipated threats. In short and in brief, there has been a shift across the sciences from regarding pathogens as predictable, episodic and calculable entities, to what are now more likely to be understood as less determinable and contingent matters. Microbial life is now often understood relationally, as a matter that is worked out between microbes, their hosts and their socio-environmental relations (Chapter 3). As those relations change so too do the microbes, who can now 'emerge' and 're-emerge' as a result of shifts in their situation.

This is, in part, epistemological, as biosciences move from approaches that highlight genetic and hereditary processes to more developmental and evolutionary approaches (Gilbert, 2014; Landecker, 2015). The shift involves a move from molecular determinants of behaviour to the contingencies and indeterminacies of ecological dynamics and development. It is also a belated recognition of the anthropogenic instabilities that include anything from accelerations in climate change, expanded and intensified trade and rapid habitat change. Certainly, there is more than discourse at play here. In the landscapes that we traced in Chapter 1, of growing human and non-human animal populations and the diversity and geographical distribution of mutable microbes and anti-microbial resistant genes, there is also, following Landecker (2015: 24), 'a material shift in the numbers, kinds, temporalities and capabilities' of microbes.

In short, what Caduff (2014) calls a cosmology of mutability is joined by an unstable world of fragile relations. Emergence here signals field transitions where the 'past', be it inherited through genes or based on archival interpretation, is a limited guide to the future as a result of both a change in understanding of microbial life (the history of biology) as well as recognition of the material shifts in global life, its mass and make up (the biology of history) (Landecker, 2015).

These biologies are clearly interrelated and affect one another. Indeed, emergence is undoubtedly used to provide further justification for alterations to the command and control elements of state-funded security, as dynamic microbes provide the opportunity to underpin more flexible and liberal approaches to disease management (Cooper, 2008) as well as extensions of sovereign power across state territories and animal and plant populations (Braun, 2007). Relational and complexity-based thinking within the biosciences seems in this sense to sanction a shift to market-based solutions to biosecurity governance and to public–private interventions in molecular life. These interventions can, in turn, raise the molecular stakes and change the evolutionary potential of microbial lives.

For now, it is worth noting that the discourses of emergence and emergencies with which we started this book, and the accounts of dynamic microbial situations that trouble germ theory and we discuss in greater detail in Chapter 3, are part and parcel of a particular reformulation of infectious disease. They help to re-diagram disease in ways that mirror and modify the diagrams that we discussed earlier. Just as leprosy, plague and smallpox involved a wrapping together of disease knowledge, religion, government and technique, so too do emerging infections, security and liberal states start to diagram disease in new ways. The redistribution of responsibilities and the growing acceptance of radical uncertainty (as opposed to the more comforting calculus of known risk and determinate species) are coupled to an ontology of emergence and the impending threat of emergency (with all its implications for governing society through fear).

Leaving things here would prompt us, like so many have done, to argue for something like a phase shift in the politics of life. But, the diagrammatic shifts we have started to trace are overlain and somewhat troubled by our fourth and fifth elements of biosecurity: a pathogen-based and anticipatory approach to infectious disease and risk. We turn to these not to refute what we have said so far, but to insist once again upon the multiplicity of any disease diagram, and the need to pay attention to the many resonances, or multi-layered nature, of disease practice.

The Return of the Pathogen

Despite the mutable and emergent microbes that we touched on above, biosecurity has a tendency to remain wedded to the concept of the pathogen. The latter tends to (temporarily at least) fix microbial identities and functions in order to provide convenient and opportune targets for public and animal health policy.

If security, in the broadest sense, has so far suggested a redistribution of responsibilities and agencies, then here, with a focus on pathogens, we are jarred back to more centripetal tendencies, and in particular the power or control *over* life. Indeed, in its focus on pathogens, biosecurity often re-diagrams disease as a matter of constructing and then protecting a system of spaces from which disease-causing agents and their vectors are excluded.

As such, the spaces of circulation that Foucault emphasised are brought into tension with territorial expressions (like zoning), borders, exclusions and inside/outside dichotomies (Hinchliffe et al., 2013). So, for example, in UK government terms, biosecurity is concerned with 'the *incursion* of infectious disease or disease vectors and their *impact* on farmed animals, crops, wildlife and humans' (Enticott et al., 2012: 327, emphasis added). Internationally, this geography of incursion is often played out across nation state territories or more latterly disease-free zones (Mather and Marshall, 2011). A trading body (often a state) or zone (a specifically designated disease-free area within a state) can submit the necessary paperwork to the World Trade Organisation (WTO) via the OIE (World Organisation for Animal Health) in order to verify their disease-free status, or conversely use WTO Sanitary and Phyto-Sanitary agreements to refuse trade in foodstuffs where there is significant evidence of disease risk. In other words, freedom *to* trade requires freedom *from* pathogens.

Liberal or free trade is thus made conditional upon a certain virtuosity that disallows or attempts to make absent certain pathogens. Confirmation of this pathogen-focused formatting of biosecurity is underlined by the OIE, the international body responsible for animal health. For the OIE, animal health is defined only in terms of the *absence* of notifiable pathogens. In OIE documentation, a positive version or definition of health is notable only by its absence. Health and disease tend to be conveyed through an epidemiological mapping of the whereabouts of pathogens. The implication and result is a spatial mapping of and will to segregate the virtuous (often the Global North) and the unruly (the Global South) (Law and Mol, 2008), and biosecurity is enacted as a continuous watch, or at least a series of risk-based surveillant operations, on possible border crossings and infringements that would carry pathogens across geographical thresholds and would, as a result, threaten freedom to trade.

It is reasonable to say that a good deal of biosecurity in the food and farming sectors, in particular, is framed by and shored up or justified in the last instance on this freedom to trade status of a region or state. Avian influenza, foot and mouth, BSE and so on are examples of trade-limiting diseases, where the presence or otherwise of a pathogen enacts a legally binding condition and interdiction. The absence of notifiable pathogenic organisms is the passport to trade and the pre-requisite of food movements more generally. It underlines the regulation of food practices from farm to fork. In short, disease diagrams of division and interdiction recur in the biosecurity disease situation.

Anticipatory Logics

Our final element of biosecurity concerns the future-oriented nature of disease management. Biosecurity is not simply reactive, it is anticipatory and aims to intervene in disease processes prior to a disease episode or event. Anticipation can involve attempts to prevent infection, take precautionary action should disease risk be identified, survey and prepare for disease, and even pre-empt any predicted disease threats (Anderson, 2010; Cooper, 2006). We will briefly introduce each of these in turn and in doing so link them to our emerging picture of a biosecurity disease diagram.

First, prevention has, characteristically, at least from the nineteenth century onwards, been framed through calculations of probabilities of a disease event. These risk calculations take a known outcome (disease outbreak) and a well-documented series of occurrences, in order to generate a likelihood of future episodes and so justify public health interventions. Seasonal winter flu in people is a good case – classically regarded as a well-known disease system, with a calculable incidence rate, as well as known outcomes in vulnerable segments of the population, flu risk provides a strong rationale for preventive actions in the form of annual vaccinations of the vulnerable and capacity matching with known case loads in clinics and intensive care units. Benefits and costs can be balanced. Preventive public health planning seems paradigmatic, in that sense, of the installation of security around the random element inherent in a population of living beings 'so as to optimise a state of life' (Foucault, 2004: 246). This optimisation relies 'on archival knowledge on the timing and location of outbreaks to design effective interventions' (Lakoff, 2008: 40).

Second, this taming of chance has informed the ways in which institutions are expected to operate as risk averse and in a precautionary fashion. The result can be not only a will to prevent disease but also a pursuit of hypertrophic security in the form of disease freedom.

This precaution is amplified by the organisational imperatives to act on dangers that may not yet have materialised. If danger exists, as legal theorist Francois Ewald puts it:

> ...it exists in a virtual state before being actualised in an offense, injury, or accident. This entails the further assumption that the responsible institutions are guilty if they do not detect the presence, or actuality, of a danger *before* it is realised (Ewald, 1993: 221–222, emphasis added).

In other words, institutions are involved in anticipating danger, and perform their responsibility through clear and auditable acts of disease detection and/or the anticipation of biological virtuality. Furthermore, this precaution is amplified by an 'anticipation of retrospection' (Miyazaki, 2003: 259) or what Caduff (2008, 2014) notes is an approach to a future which may well involve being called to account for the actions that were or were not taken to reduce disease risk. The inference is that security does not just involve intervention and calculation (or normalisation as it was termed in Table 2.1), it also requires anticipation and, in this case, returns us to interdiction and legality (division). Anticipation of known and unknown threats exerts pressure on food producers, retailers and regulators, who either assert their sanitary agency over the living processes they organise or specify contracts and legal responsibilities such that, should the worst happen, guilt cannot be linked to the actions of the institution or organisation. The result of this precautionary landscape is a territorial and Euclidean interpretation of biosecurity with a networked performance of distributed accountability and responsibility.

Third, as may now be clear, these approaches to health planning and prudence have, in recent decades, undergone challenges from liberal approaches to governance and from evolutionary understandings of mutable microbes (see above). The non-probabilistic nature of microbial behaviour, the erosion of state regulation, as well as the perceived hyper-connectivity of global communication and travel, have ushered in a shift in anticipatory logic from prevention of known risks to preparedness for unknown (but seemingly inevitable) disease events (Collier and Lakoff, 2008a, 2008b; Lakoff, 2008). The latter is the ground upon which to legitimise a state of continuous vigilance and readiness, and an infrastructural investment in watchful and primed surveillance (WHO, 2007).

The result is a tendency to draw future disease threats into the present. Enhanced surveillance, for example, tends to combine a view on the mutable excesses of disease (the disease emergency to come), its imminent emergence, with an investment in emergency measures that need to be taken to counter its worst effects. In other words, mutable molecules are rendered as radically indeterminate (and dangerous) at the same time as authoritative social power and control are normalised (Caduff, 2014). Indeed, a future–present catastrophe, or breakdown augured by surveillance systems, seems to be an invitation to reinforce established routines and powers. Surveillance systems like the WHO's *FluNet*, as well as modelling platforms like *Epicast* (the National Infrastructure Simulation and Analysis Centre's epidemic simulations), for example, 'do not so much surveil the future as construct the present as catastrophic' (Thomas, 2014: 288) and in need of (sovereign) control.

This vision is reproduced in any number of infectious disease maps that highlight connectivity and flow. Coupled with texts that tend to oppose a world of virtuous European/North American human control with non-Western contingency or interspecies intimacy, these maps present the horror of microbial movement from East to West and South to North. The result is the impression of an 'easy, smooth transmission of microbes, goods and people across various borders and through various scales, smoothing over local disconnection, friction and delays to create easy-to-digest maps of infection' (Thomas, 2014: 291). Their presentation of 'real-time' events or a 'continual, ever-present unfolding' implies not only speed (the immediate conversion of signals and symptoms into data) but also a singular view of space as a shrinking surface over which objects and actors move with increasing ease (Figure 2.1).

There is a cartographic tendency in these images to reduce disease to pathogens, and to abstract pathogens from their relational spaces. In other words, despite the ontological shift to mutability informing a shift to preparedness, the latter's own technologies of visualisation and mapping seem to re-diagram disease as a matter of pathogens *in* space. The result is a tendency to seek interdiction in the face of too much connectivity and a will to enclose spaces in order to prevent movement. We are, in that sense, drawn back to a world where disease is diagrammed through sovereignty and discipline as well as security.

The same can be said of pre-emption, our fourth and final anticipatory logic. If prevention acts in response to a known threat, and preparedness involves readiness in a world of biological virtuality for an as yet unknown event, then pre-emption involves the speculative attempt to counter possible future threats by mobilising innovation (Cooper, 2006). It involves immersion in the conditions of emergence to the point of actualising the future in advance. A key example here would be the gain-in-function experiments conducted on the highly pathogenic avian influenza virus H5N1 in order to speed up changes to transmissibility in mammals. The result was the production, in the laboratory, of an influenza virus that was not only pathogenic, but also well adapted to and therefore transmissible between ferret models (Fouchier et al., 2012). This pre-emption of mutations that could occur 'in the wild' enabled a molecular characterisation of those future viruses. The resulting biological information, in the form of a genetic code, could, protagonists argued, provide the antigenic information for the manufacture of an effective vaccine well ahead of the onset of a pandemic. Vaccine manufacture can take anything up to 6 to 10 months following characterisation of a new flu virus, making the pre-emption of the arguably soon to naturally occur antigenic structure obviously attractive.

The irony, of course, was that as virologists produced 'information value' through these gains in function experiments, so too was the scope of security expanded. Not only did the world need to deal with mutable viruses, it now had to find ways of securing the information or code that related to the molecular structure of highly pathogenic viruses generated within a laboratory setting. And, just to add to the issue, securing information value needed to attend to the intellectual property claims on that information, not only from scientists but also those supplying wild-type viruses. The 2007 standoff between the Indonesian Government and the World Health Organisation is a case in point where disease is pulled and shaped by various diagrams. The submission of live avian influenza virus samples to international reference laboratories without any undertaking that donor countries (who were suffering the brunt of the disease) would receive

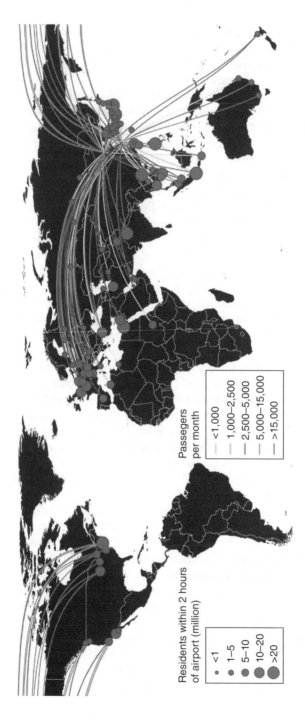

Residents within 2 hours
of airport (million)

· <1
· 1–5
● 5–10
● 10–20
● >20

Passegers
per month

— <1,000
— 1,000–2,500
— 2,500–5,000
— 5,000–15,000
— >15,000

Figure 2.1 A map of disease risk linked to air passenger movements and human population centres, an example of real-time surveillance producing a catastrophic present. (Butler, 2013. Reproduced with permission of Nature Publishing Group.).

compensation or materially benefit from the possible vaccines or other therapies that ensued from viral characterisation, produced a debate over 'viral sovereignty', intellectual property rights and the uneven distribution of benefits and costs of a so-called global health. Infectious disease is in this sense diagrammed as a matter for anticipatory action, but also an issue that requires attention to be given to uneven resources and structural inequality.

Within each of these anticipatory approaches we can see disease being diagrammed in multiple if overlapping ways. Prevention and precaution tend to map onto pathogen-focused approaches, while preparedness and pre-emption assume a more mutable world. Nevertheless, as we have been keen to point out, within each of these approaches there are tendencies that pull the disease diagrams in different directions. How these directions are coordinated, resolved or otherwise will depend, we will argue, on the disease situation.

Biosecurity is clearly a composite activity in which various diagrams of disease tend to push and pull microbes, hosts and the socio-economies of disease in a number of directions. Territories, flows, real-time as well as futures are all drawn into its remit. How these various spaces and times, as well as bodies and processes, are held together, often in tension, is a key issue for empirical inquiry. This is not about one always dominant logic, that is ripe for ideology critique, but is rather a matter of being sensitised to the multiple logics and diagrams that make a biosecurity situation. In the conclusion we amplify these points before returning to the issue of a situation which is constituted, amongst other things, by the meeting up of various disease diagrams.

Conclusions

We have identified five broad components of biosecurity – security, liberalism, mutable microbes, pathogen-based and anticipatory approaches to disease. Taken together they underline that biosecurity cannot be reduced to one disease diagram or form of power. Biosecurity is composite and exists in tension with itself. In all of these components, but most notably once they are combined together, biosecurity seems caught between the 'brighter line' (Leduc, 2010) of demarcation and interdiction, and the muddier logics of security. While the latter troubles any hard and fast distinctions between good and evil, the permitted and the forbidden, the former speaks to a will to control and eliminate, to demarcate a virtuous space of good practice from spaces that are socially and microbiologically contingent.

Here it is useful to reiterate two points. First, the diagrams of disease with which we started this chapter (exclusion, inclusion and normalisation) are clearly multi-layered and far from being mutually exclusive. In fact, they operate in tandem with one another, despite their potential disagreements or the antagonisms that they would seem to suggest. In this sense, disease is always, it seems, plagued by spatial tensions. Second, biosecurity replays many of these spatial tensions. While the rise of security and liberalism, and even the shift to more evolutionary understandings of microbes, suggest on first blush that infectious disease is being diagrammed as a matter for security rather than discipline or sovereignty, there are countervailing elements to biosecurity. Pathogens and territories remain as matters to be kept apart. Notifiable disease becomes a matter of sovereignty and interdiction. Likewise, anticipation can involve centripetal tendencies and emphasise a misplaced human power over life. Some of the temporalities may have changed (pre-emption and preparedness would be the most frequently mentioned here), some of the modes and some of the targets have altered (as pathogens shift from being self-identical to emergent), but here concerns with microbial agents and interdictions simultaneously draw us back to a diagram of exclusion. An outside remains.

The resulting mix of diagrams makes biosecurity an ongoing compromise between economic circulation, the regulation of 'surplus' life and the organisational requirement to make life safe. How the various diagrams are coordinated – its ontological politics as Mol (1999) would call it – is a matter for empirical investigation and helps to constitute what we are calling disease situations (Chapter 3). For now, it is worth noting that the tensions can produce some interesting, and even, paradoxical situations. For example, the problem with sanitary and territorialised versions of biosecurity that are set up in pathogen-focused and anticipatory settings is that life can seemingly be made safe while disqualifying the very processes that make it responsive to and able to deal with inevitable perturbations. The result can be a reduction of life to its barest of conditions, a temporary triumph of mere life in order to suppress more life (Chapters 4 and 5). Here it may well be the very affirmation of agency and the assumption of abundant capacities of deliberate human action, coupled with emergent biosocial phenomenon like infectious disease, that is the cause for concern (Lee and Motzkau, 2013).

Diagrams are not representations, they perform disease and have effects. The diagram 'continually churns up matter' and affects change (Deleuze, 1999: 35). How this churning is affected in practice becomes an empirical question and one that we relate in Chapter 3 and in Part II to specific disease situations. Does, for example, the continuing exteriorisation of disease produce the kinds of paradoxical effects that we have hinted at in this chapter? Do countervailing tendencies of shifting responsibilities and of modulation reduce these effects or, indeed, create new kinds of problem? Our point, for now, is to underline the need to work empirically in order to trace disease diagrams and attend to how tensions are resolved or lived with in practice. What kinds of issues, for example, tend to pull a diagram one way rather than another? And with what kinds of effect? In order to start this process we turn now to the conceptual and methodological issues that are involved in studying disease situations.

References

Anderson, B. 2010. Preemption, precaution, preparedness: Anticipatory action and future geographies. *Progress in Human Geography*, 34, 777–798.

Anderson, B. & Adey, P. 2011. Affect and security: Exercising emergency in UK civil contingencies. *Environment and Planning D: Society and Space*, 29, 1092–1109.

Anderson, W. & Mackay, I.R. 2014. *Intolerant Bodies: A Short History of Autoimmunity*. Baltimore, MD: Johns Hopkins University Press.

Artaud, A. 1970. Theatre and the Plague. *In*: Artaud, A. (ed.), *The Theatre and its Double*. Richmond, UK: Alma Classics.

Barker, K. 2010. Biosecure citizenship: Politicising symbiotic associations and the construction of biological threat. *Transactions of the Institute of British Geographers*, 35, 350–363.

Braun, B. 2007. Biopolitics and the molecularization of life. *Cultural Geographies*, 14, 6–28.

Butler, D. (2013). Mapping the H7N9 outbreaks. *Nature*, doi:10.1038/nature.2013.12863.

Caduff, C. 2008. Anticipations of biosecurity. *In*: Lakoff, A. & Collier, S.J. (eds), *Biosecurity Interventions: Global Health and Security in Question*. New York: Columbia/SSRC.

Caduff C. 2014. Pandemic prophecy: Or, how to have faith in reason. *Current Anthropology*, 55, 296–305.

Canguilhem, G. 1991 (1966). *The Normal and the Pathological*. New York: Zone Books.

Chien, Y.-J. 2012. How did international agencies perceive the avian influenza problem? The adoption and manufacture of the 'One World, One Health' framework'. *Sociology of Health and Illness*, 35, 213–226.

Collier, S.J. & Lakoff, A. 2008a. Distributed preparedness: The spatial logic of domestic security in the United States. *Environment and Planning D: Society and Space*, 26, 7–28.

Collier, S.J. & Lakoff, A. 2008b. The problem of securing health. *In*: Collier, S.J. & Lakoff, A. (eds) *Biosecurity Interventions: Global Health and Security in Question*. New York: Columbia University Press/SSRC.

Cooper, M. 2006. Pre-empting emergence: The biological turn in the war on terror. *Theory, Culture and Society*, 23, 113–135.

Cooper, M. 2008. *Life as Surplus: Biotechnology and Capitalism in the Neoliberal Order*. Seattle: University of Washington Press.

Deleuze, G. 1999. *Foucault*. London: Continuum.

Dillon, M. & Lobo-Guerrero, L. 2008. Biopolitics of security in the 21st century. *Review of International Studies*, 34, 265–292.

Enticott, G., Franklin, A. & Van Winden, S. 2012. Biosecurity and food security: Spatial strategies for combating bovine tuberculosis in the UK. *Geographical Journal*, 178, 327–337.

Ewald, F. 1993. Two infinities of risk. *In*: Massumi, B. (ed.), *The Politics of Everyday Fear*. Minneapolis: University of Minnesota Press.

Fischer, M.M.J. 2012. On metaphor: Reciprocity and immunity. *Cultural Antrhopology*, 27, 144–152.

Fleck, L. 1979. *Genesis and Development of Scientific Fact*. Chicago: University of Chicago Press.

Foucault, M. 1977. *Discipline and Punish: The Birth of the Prison*. Harmondsworth, UK: Penguin.

Foucault, M. 1981. *The History of Sexuality*, vol. 1: *An Introduction*. Harmondsworth, UK: Penguin.

Foucault, M. 2004. *Society must be Defended*. Harmondsworth, UK: Penguin.

Foucault, M. 2007. *Security, Territory, Population: Lectures at the College de France 1977–78*. London: Palgrave Macmillan.

Fouchier, R.A.M., Garcia-Sastre, A., Kawaoka, Y. & Others, A. 2012. Pause on avian flu transmission studies. *Nature*, 481, 443.

Gilbert, S.F. 2014. Symbiosis as a way of life: The dependent co-origination of the body. *Journal of Biosciences*, 39, 201–209.

Gradmann, C. 2009. *Laboratory Disease: Robert Koch's Medical Bacteriology*. Baltimore, MD: Johns Hopkins University Press.

Hardt, M. & Negri, A. 2005. *Multitude*. London: Penguin.

Hinchliffe, S. 2007. *Geographies of Nature: Societies, Environments, Ecologies*. London: Sage.

Hinchliffe, S. & Bingham, N. 2008. Securing life – the emerging practices of biosecurity. *Environment and Planning A*, 40, 1534–1551.

Hinchliffe, S., Allen, J., Lavau, S., Bingham, N. & Carter, S. 2013. Biosecurity and the topologies of infected life: From borderlines to borderlands. *Transactions of the Institute of British Geographers*, 38, 531–543.

Keck, F. 2008. From mad cow disease to bird flu: Transformations of food safety in France. *In:* Collier, S.J. & Lakoff, A. (eds), *Biosecurity Interventions: Global Health and Security in Question*. New York: Columbia University/SSRC.

King, N.B. 2002. Security, disease, commerce: Ideologies of post-colonial global health. *Social Studies of Science*, 32, 763–789.

Lakoff, A. 2008. From population to vital system: National security and the changing object of public health. *In:* Collier, S.J. & Lakoff, A. (eds), *Biosecurity Interventions: Global Health and Security in Question*. New York: Continuum.

Landecker, H. 2015. Antibiotic resistance and the biology of history. *Body and Society*, doi: 10.1177/1357034X14561341.

Latour, B. 1983. Give me a laboratory and I will raise the world. *In:* Knorr-Cetina, K.D. & Mulkay, M. (eds), *Science Observed. Perspectives on the Social Study of Science*. London: Sage.

Latour, B. 1988. *The Pasteurisation of France*. Cambridge, MA: Harvard University Press.

Latour, B. 1999. *Pandora's Hope: Essays on the Reality of Science Studies*. Cambridge, MA: Harvard University Press.

Law, J. (ed.) 1991. *A Sociology of Monsters*. London: Routledge.

Law, J. & Mol, A. 2008. Globalisation in practice: On the politics of boiling pigswill. *Geoforum*, 39, 133–143.

Leach, M. & Dry, S. 2010. Epidemic narratives. *In:* Leach, M. & Dry, S. (eds), *Epidemics: Science, Governance and Social Justice*. London: Earthscan.

Leduc, J.W. 2010. Preface. *In: Sequence-based Classification of Select Agents: A Brighter Line*. Washington DC: National Academies of Sciences.

Lee, N. & Motzkau, J. 2013. Varieties of biosocial imagination: Reframing responses to climate change and antibiotic resistance. *Science, Technology and Human Values*, 38, 447–469.

Lentzos, F. & Rose, N. 2009. Governing insecurity: Contingency planning, protection, resilience. *Economy and Society*, 38, 230–254.

Mather, C. & Marshall, A. 2011. Biosecurity's unruly spaces. *The Geographical Journal*, 177, 300–310.

Methot, P.-O. & Alizon, S. 2014. What is a pathogen? Towards a process view of host–parasite interactions. *Virulence*, 5, 775–785.

Methot, P.-O. & Alizon, S. 2015. Emerging disease and the evolution of virulence: The case of the 1918–1919 influenza pandemic. *History, Philosophy and Theory of the Life Sciences*, 7, 93–130.

Miyazaki, H. 2003. The temporalities of the market. *American Anthropologist*, 105, 255–265.

Mol, A. 1999. Ontological politics, a word and some questions. *In*: Law, J. & Hassard, J. (eds), *Actor Network Theory and After*. Oxford and Keele: Blackwell/Sociological Review.

Mol, A. 2002. *The Body Multiple: Ontology in Medical Practice*. Durham, NC: Duke University Press.

Paxson, H. 2008. Post-pasteurian cultures: The micobiopolitics of raw-milk cheese in the United States. *Current Anthropology*, 23, 15–47.

Schrader, A. 2010. Responding to *Pfiesteria piscicida*: Phantomatic ontologies, indeterminacy, and responsibility in toxic microbiology. *Social Studies of Science*, 40, 275–306.

Scoones, I. 2010. Fighting the flu: Risk, uncertainty and surveillance. *In*: Dry, S. & Leach, M. (eds), *Epidemics: Science, Governance and Social Justice*. London: Earthscan.

Scoones, I. & Forster, P. 2011. Unpacking the international responses to avian influenza: Science, policy and politics. *In*: Scoones, I. (ed.), *Avian Influenza*. London: Earthscan.

Thacker, E. 2009. The shadows of atheology: Epidemics, power and life after Foucault. *Theory, Culture and Society*, 26, 134–152.

Thomas, L. 2014. Pandemics of the future: Disease surveillance in real time. *Surveillance and Society*, 12, 287–300.

Tomlinson, I. & Potter, C. 2010. Too little, too late? Science, policy and Dutch Elm Disease in the UK. *Journal of Historical Geography*, 36, 121–131.

Wolfe, C. 2013. *Before the Law: Humans and Other Animals in a Biopolitical Frame*. Chicago and London: The University of Chicago Press.

World Health Organisation 2007. A safer future: Global public health security in the 21st century. *World Health Report*. Geneva, Switzerland: World Health Organisation.

Yates-Doerr, E. 2015. The world in a box? Food security, edible insects, and 'One World, One Health' collaboration. *Social Science and Medicine*, 129, 106–112. doi: 10.1016/j.socscimed.2014.06.020.

Chapter Three
Reconfiguring Disease Situations

As we saw in Chapter 2, contemporary approaches to biosecurity involve a mixed bag of techniques and practices. In the terms we have used, they tend to incorporate more than one disease diagram. Exclusion, inclusion and normalisation co-exist, and there is, we have argued, a mix of spatialities in play. The territorial and trade-related demands of disease-free territories, for example, often vie for attention with the will to expand and accelerate flows of goods and capital. Emergent and relational accounts of dynamic host–microbe relations are in tension with pathogen-centred accounts of disease causation and aticipation. These mixed orderings are in tension with one another. How these tensions are managed characterises, in part at least, what we term disease *situations*.

In using the term 'situation' we mean to emphasise two things. Situations are *first* a meeting up of the various spatial and temporal orderings that constitute various disease diagrams. Here exclusion and inclusion can clash or rub up against one another, and meet dynamic microbes, ballooning livestock populations and tight economic margins. Second, in this awkward mix, situations can spark or create new issues, problems or matters. This rubbing together of various matters and orderings mean that situations may be 'charged', or have a potentiality that can generate events, prompt a shift in attention and new actions. Situations are in that sense real and existing manifestations of multiple processes

Pathological Lives: Disease, Space and Biopolitics, First Edition. Steve Hinchliffe, Nick Bingham, John Allen and Simon Carter.
© 2017 John Wiley & Sons, Ltd. Published 2017 by John Wiley & Sons, Ltd.

that also have a power to force thought. This power is always 'a virtual one' that 'has to be actualised' (Stengers, 2005b: 185). In other words, situations offer the potential for rethinking the emerging disease emergency. In this chapter we set up situations in more detail and demonstrate their methodological value for studying emerging infections.

In Chapter 2 we pointed out that diagrams are not set in stone. Rather, they are continuously being redrawn as ideas and practices shift or diverge in the course of conduct. Indeed, we argued that the ways in which diagrams interrelate and are enacted or performed is something of an open question, and one for which empirical analysis is needed. There may, for example, be practices that change the diagram or alter the ways in which disease is addressed or reduced. In this sense, diagrams are resources for understanding and analysis, but can and will play out differently within different empirical set-ups. How to study such set-ups is the focus of this chapter. In addressing this question and by interrogating how disease diagrams are altered or interfere with one another, we turn to this composite notion of disease situations. In doing so, our broader aim is to ask whether a reconfiguration of disease diagrams that may occur within particular disease situations can help us to reframe the emerging infectious disease emergency.

This chapter lays out some of the resources for and means of studying disease situations. In doing so, we prepare the groundwork for Part II of *Pathological Lives*, in which we look at particular disease situations (or aspects of disease situations) in more empirical detail. The current chapter is split into three subsections. We start with a discussion of disease situations, emphasising two key aspects. First, the shift from contamination to configurational understandings of disease and, second, the ways in which a situation exceeds and yet is intricate with individual disease sites. We then go on to develop an alternative account of infectious disease. It is an account that displaces pathogens *per se* and invites us to look for the production of pathogenicity. This shift of focus allows us to trace some of the key components of disease situations, or, in other words, what makes a disease situation turn pathological? What converts microbes into pathogens, and what gives them pathogenicity? Answering these questions involves bringing microbiological understandings of the societies of microbes together with a social science understanding of the societies of people, animals and economies. Once we have set up this social understanding of pathogenesis, we finally start to conceptualise the ways in which all the ingredients of a disease situation are held together. How to conceptualise and imagine the components of a situation? Here we expand on resources from spatial theory, and in particular the topological thinking that allows us to provide an analytical lens on disease situations. We end the chapter with a discussion of the methodological implications and approaches for studying situations. What resources exist for social scientists to study such disparate and complicated spaces?

Disease Situations

Disease situations are, we argued in Chapter 1, meeting places where various actors, bodies, species, business pressures, microbes, technologies and so on are organised, brought together or held apart and worked upon. They are constituted from and constitutive of a mix of disease diagrams. They are distinct from sites in the sense that these meetings are a constellation of matters that may not, in some senses, all be physically present at the site.

The term situation provides us with at least two advantages. First, the term alerts us to the intra-actions between, in epidemiological terms, environments, hosts and microbes. It reminds us that pathogenicity is an outcome of these intra-actions rather than the mere location and presence of a microbe. While microbial presence is clearly vital, it is also of course insufficient on its own to explain or cause disease. Disease or the production of conditions for replication and transmission, or indeed the often secondary toxic effects or immunological responses that may be associated with the disease, are dependent on a range of factors or variables including for example species type, living conditions, presence of other microbes and so on. But more than this, and second, a situation exceeds individual sites. It is a combination or a relating together of matters that can be distributed across many sites, institutions and practices, and involve all manner of people, other species, materials and processes. A situation in this sense demands that we trace or reason back from an infected premise or site and in turn focus on relations that matter. Following Barry (2013: 93), this process can also involve a more transformative 'abductive logic' through which situations can be built and attention directed 'towards the importance of a wider political controversy'.

To illustrate what a situation is and what it means to abduct from a site, we can use an example. An outbreak of avian influenza on a relatively 'bio-secure' indoor duck breeding farm near Driffield, East Yorkshire, England took place in late 2014. The alarm was raised by a veterinary surgeon working out of a private practice, after staff at the farm had noticed a handful of unexplained fatalities and a drop off in egg production. The government's Animal and Plant Health Agency (APHA) were called, and blood as well as throat and cloacal swabs were taken from a sample of the housed birds. The virus was initially difficult to identify (for an account of viral surveillance, see Chapter 8). The ducks had several other illnesses (and indeed may not have come to the notice of the farm manager and veterinarian had there not been other, so-called intercurrent, conditions), but molecular and virological work at APHA's European Reference Laboratory suggested that the birds had been infected with an H5 avian influenza virus (the H refers to haemagglutinin, a protein that enables a virus particle to enter host cells. Sixteen types have been identified in birds, each numbered, H1, H2 and so on). The H5 molecule, along with H7, is a protein of concern as it is well adapted to birds (so highly contagious) and often deadly in galliformes (the order that includes chickens and turkeys). These virus strains may also be transmissible to people, and once in people are associated with high mortality rates. Moreover, the gene that codes for this protein is also highly promiscuous and mutable, leading some to worry that these virus types could lead to the next human pandemic, which effectively would involve a ramping up of the transmission rate between people.

A second key protein, neuraminidase (NA), effectively enables the virus to escape the host's cell once reproduction has occurred. Together and in intra-action the entry and exit proteins help to explain the infectivity and transmissibility of the virus. In the case of the Driffield ducks, the neuraminidase took longer to positively identify, but the N1 molecule was quickly ruled out, allowing officials to confirm that the human health implications of the outbreak were not critical. (H5N1 being the influenza group that had caused most zoonotic concern and fatalities in people over the last 18 years).

All the evidence suggested that the ducks in Yorkshire had been infected with the same strain of virus (H5N8) that was associated with recent outbreaks in Dutch chickens and German turkeys. It was a 'highly pathogenic' strain, meaning that it caused severe disease and death in chickens, a definition of pathogenicity to which we will return, and was therefore a notifiable disease that required statutory action in the form of movement bans, exclusion and surveillance zones. Further genetic sequencing of the Driffield virus revealed that it shared over 99% of its genome with the viruses in Germany and The Netherlands. Furthermore, the H5N8 strain was similar to one that had been circulating in China, Japan and Korea for the last four years and had been associated with outbreaks in Korea and Japan within the last 10 months.

The coincidental infection of three different domestic species in three European countries in a matter of days and weeks suggested a common cause. In the parlance of epidemiologists, the similarity in the virus strain indicated that the initial source (or incursion) of infection was the same for each of the infected premises. The default explanation for such a pattern of disease, and indeed the spread of H5N8 from East Asia, was that the pathogen had been circulating in wild bird populations and had moved across Europe at a time when waterfowl in particular were migrating westwards in order to over winter in milder maritime climates. The infected premises were all on a wild bird migratory pathway linking Eurasia (as far east as Western Siberia) and Northwest Europe. The positive testing of a wild bird (teal) in Germany for H5N8 seemed to add some weight to the wild bird hypothesis, even if the sampling process was hardly robust.

And yet, the Yorkshire farm was regarded as bio-secure with respect to wild bird incursion and was at a location where few wild birds were detected by expert ornithologists. Moreover, there were plenty of other premises nearby, including outdoor poultry farms, where biosecurity was less evident and which had not reported disease. In chickens it would not require a veterinarian and a virology laboratory to detect the presence of the virus, as birds would normally die en masse within a day or so of the virus reaching the flock. Likewise, the 'smoking gun' of the infected teal did not explain the transmission from East Asia. The bird populations of Asia and Eurasia are normally regarded as separate populations, meaning that there would have had to have been a series of jumps necessary to move the virus halfway around the world (Lee et al., 2015).

There remained, in other words, considerable uncertainty with respect to the wild bird hypothesis. It *was* possible that intra-population mixing had occurred in shared feeding areas in Eurasia, which had resulted in eastward and westward movement of the virus. Likewise, spillover from a wild bird reservoir might have been part of the explanation for the Driffield duck case, and for that matter all of the indoor infected premises in northern Europe. Even if infected wild birds had not had access to the Driffield ducks, there was a chance that transmission could have been indirect. The virus can remain viable in fomites (in various materials like excreta and water) and could circulate on people's shoes, on vehicles, on vectors like rats. But the evidence of the actual pathway of infection was at best incomplete (European Food Safety Authority, 2014).

In the Driffield case, a question remained as to why a seemingly more bio-secure site had suffered a breakdown when there were 208 premises within a 10 km radius with susceptible poultry stock and often with much less rigorous or even absent biosecurity (Animal and Plant Health Agency, 2014). The same puzzle was to occur in Spring 2015 when the US experienced its worst ever avian influenza epizootic, with H5N8, similar to the one that infected the Driffield ducks, and then a related virus H5N2, devastating the turkey, chicken broiler and layer industry. The result was the culling of 50 million chickens and turkeys across 21 US states (USDA, 2015). Again, the 219 reported detections were nearly all located at commercial, large farms with high levels of biosecurity. Only 12 'back yard' or open-air farms reported the disease across the US. In a similar vein, evidence from the 2006 outbreak and the endemic nature of H5N1 in Egypt has started to implicate the commercial sector in the production and persistence of the disease (Dixon, 2015; Hinchliffe and Bingham, 2008).

Given this pattern of affected sites, it was possible, at least, that wild birds were something of a distraction, though this is not the tenor of most official epidemiology reports. Large avian populations and high stocking densities of commercial premises, plus the number and frequency of inter-site connections, may well provide an alternative focus for attention. Rather than spillover from wild birds, and the absence or otherwise of bio-secure borders, the 'infectability' of the industry (for more on this see Chapter 4) and spill-back into the wild world of circulating influenza viruses (WHO, 2015) may be causes for concern.

Indeed, in this sense we could say that the disease situation was not only virological and ornithological. There were socio-ecological and economic aspects to consider. The company that owned the farm in Yorkshire was a large multinational that bred and finished ducks in England, Germany and China. 'Cherry Valley' was one of the biggest producers of ducks in the UK, rearing 7 million ducks a year, and operating a breeding pyramid that involved numerous premises (from grandparent breeding or elite stock, to layer or multiplier units and finishing stock for slaughter), and the movement of stock, eggs, feed and staff. The Driffield ducks were part of the mid-section of the breeding pyramid, supplying eggs and eventually ducklings to contracted and subcontracted premises throughout the UK, who grew them on for eventual slaughter and sale. Meanwhile, this operation was dwarfed by its Chinese parent company, which was estimated to rear 2.8 *billion* ducks per year. Investigations were clearly needed into the economic and material relations between domestic flocks within the UK, across Europe and with Asia. The key point here is that the infected site was linked to many other sites within a corporate structure. As ducks (from the wildfowl or Anseriformes order) tend to display few if any symptoms once infected with influenza viruses, it is possible that the virus was circulating undetected within the company structure. Whether or not this was the case, the comings and goings of stock and materials, as well as staff within and between sites, merited further investigation. They provided another set of issues that related to the disease situation.

A final point to emphasise is that the disease situation is not just a matter of looking for alternative infection pathways. Infection is but one component of an epidemiological triangle that includes hosts and environments. So the situation also encompassed the industry itself, the throughput of birds, their living bodies and tissues, the densities and intensities of production. The industry affected, one could argue, everything from the immune responses of the ducks to the incentives or otherwise of farmers to report sick birds. Poultry farming is big business, directly worth £3.3 billion to the UK economy (British Poultry Council, 2014). The loss of England and Wales's disease-free status, once HPAI (Highly Pathogenic Avian Influenza) had been confirmed, resulted in a fall of trade across the poultry sector to other EU member states and to other (third-party) trade partners. The generation of economic surplus and the circulation of currency within and outside the nation states where H5N8 had been detected was clearly part and parcel of the disease situation.

The point is not to exhaust this case, it is rather to emphasise that the disease situation is an assemblage of numerous matters, concerns and relations. It starts to bring together the trials of laboratory science, diagnosis and molecular biology, with ancient migratory movements of birds, climate and weather, duck physiology, poultry farming, labour and stock sharing, stocking rates, profitability, corporate links, international trade and so on. In sum, the highly pathogenic avian influenza *situation* is more than a matter of an outbreak *site*. It involves what the anthropologist Celia Lowe (2010) refers to as the viral cloud, the suite of issues touched by and shaping the avian influenza virus. We take situations to be similar in intent to clouds, though we like the more grounded feel of 'situation'. The term also evokes, for us, a sense of the predicament or political openings that the disease event presents. As we noted at the start of the chapter, a situation is something that seems to invite further action, to shift attention and force thought.

To make sense of this and similar disease situations we have already made use of two supplements to the notion of an infected premise or site. First there is the need to consider the infectability or pathogenicity of a situation and second the logic that involves both reasoning back from a site and the linking together of sites in order to recompose that situation and offer new insights. To say something more about each of these, we will introduce configurational understandings of disease, and second, a social theory of situations.

Contamination and Configuration

Infectious disease, as we have seen, is often explained or diagrammed as a matter of untoward contact with an outside (a wild bird, a miscreant producer) *source*. Stopping crossover of microbes into a population of domestic animals is the first line of defence, something that is followed by the containment of any infected premises in order to prevent onward transmission, or *spread*. This geometry of source and spread leads to an oft-stated aim of managing the interfaces between wild and domestic populations, between people and animals, and among people if a virus succeeds in crossing species 'boundaries' and becomes infectious in its new host.

Following from this geometry of spread, there is a standard view that increasing centralisation and regulation of livestock agriculture would help to police these interfaces, reduce incursion and prevent spread. For example:

> The global trend of centralising livestock production for meat, dairy, and poultry on larger farms, as opposed to smaller more decentralised holdings, might provide an opportunity to introduce standards and strategies for surveillance for animal populations that would have been previously more difficult to regulate and maintain. (McCloskey et al., 2014: 1007).

This vision of security involves divisions between secure and contingent life, between domesticity and wildness, between disease-free and risky stock. The intention is clear: to enclose domestic life, centralise its organisation, include it in an enhanced surveillance practice and regulate its interactions with the outside world to such an extent that we start to hear the phrases 'disease free' and 'high health' used to describe those portions of the livestock sector and those nation states that conform to this way of farming. McCloskey and colleagues link this to the production of *One World One Health*, or the integration of human and veterinary public health with the shared end of a reduction in the infectious disease burden across all species (Craddock and Hinchliffe, 2015). In passing, it is worth noting that this one health vision is dependent on the construction of not one but two worlds – a world of regulated and highly commercialised indoor agriculture, and another world characterised by (or troped as) unregulated interspecies mixing. The virtuous, pathogen-free world that is the ideal of enclosed and surveillant livestock economies is contrasted to a pathologised world of contingency. The latter is marked by human, domestic and wildlife interactions and the microbiological crossings that are associated with infections and disease (Hinchliffe, 2015).

Seen in this form, biosecurity merely represents a rebranding of the centuries-old battle with disease (see Chapter 2); the struggle to separate diseased from healthy life, to contain infectious outbreaks and to police the flow and movement of anything potentially threatening to life. As the cultural and political theorist Angela Mitropoulous (2012) characterises it, this is a typical (and in her terms a masculine) response to the contingencies of contact. Contact, or the with-ness of touch, always provides the possibility for the degradation as well as the enhancement of life, such that touch increasingly becomes subject to regulation, a biopolitics that, as we have seen, always risks disqualifying that which makes life liveable. From sexual partners to global trade, from migration of people to blood transfusions, there is always a fraught relation between those in contact. This risky contact, she argues, is characteristically governed and managed in a particular way. *Movements* are progressively translated to regulated circulation, *contact* to regulated exchange and *communication* to commerce. Essential contact, in short, is managed, cleaned up, economised and sanitised. Living processes are reduced to the bare essentials of exchange under conditions that are sanctioned and regulated.

The understanding of disease that underpins this management of inter-species, intra-species and inter-human crossings is characterised as *contagionism*, the belief that infection and spread is predominantly a matter of unregulated contact or *contamination* of the healthy by the unhealthy or sick. For the historian of medicine Charles Rosenberg (1992: 293), contamination is one of two fundamental modes or styles of explanation for epidemics that have been 'conceptually "available" since classical antiquity'. It reduces disease to infection and to physical contact, and is more latterly bolstered of course by germ theory, which can reduce infectious disease to the transmission of microbes (see Chapter 2).

Rosenberg's second broad mode of explanation is 'configuration', or a less reductive understanding of disease. Here microbes are but one component of a complex and indeterminate interspecies process. Rosenberg argues that both modes and their supporters have vied for attention over a long period, and remain active. For example, in the nineteenth century, contagionists battled with anti-contagionists over explanations for, and responses to, diseases like cholera and yellow fever. The anti-contagionist movement was buoyed in part by commercial interests that opposed fifteenth-century quarantine and other contagionist apparatus for disease control on the grounds that such devices were often costly and ineffective. They used conflicting evidence on epidemics, the cases of contaminated people who did not develop disease and diseased people who could not be contaminated, to suggest that infection was not simply a matter of physical contact (Rosenberg, 2009; Ackerknecht, 1948). These anti-contagionists were later joined by the contingent contagionists, a mid-nineteenth-century consensus that accepted 'the transferability of disease from the sick to the well, but only in circumscribed situations' (Rosenberg, 1992: 298). Both tended to lose ground as 'germ theory helped swing medical opinion toward contamination in its modern, laboratory-oriented guise' (Rosenberg, 1992: 299).

Even so, anti-contagionist or configuration-based ideas lived on within social medicine, in clinical practice and in parts of public health. While Rosenberg limits himself to human medicine, it may be presumed that certain vets, farmers and food health practitioners also maintained an element of configuration-based explanations for epizootics and food-borne diseases, even if the prevailing trend was towards contamination-based explanations for and responses to infectious disease. For example, the politically radical nineteenth-century German pathologist Rudolf Virchow not only insisted on medicine as a social science but also argued for greater links between veterinary and medical practice (Zinsstag et al., 2011). He promoted a more holistic style of human and animal medicine, and is now regarded as an innovative forerunner of 'one health', even if the latter is now as likely as not to work with a contagionist view of infectious disease (Hinchliffe, 2015).

The point for now is that contamination as an explanation is significant in mainstream responses to emerging infections, but we should not assume that configuration or more relational approaches to understanding disease and its management have disappeared. Rather than accept a history of necessary and science-based progress, Ackerknecht and Rosenberg remind us that, first, contagionism pre-existed the science of germ theory, having ancient roots that were re-inscribed in religious texts. Germ theory tipped the balance towards an already established programme of separation or quarantine and rekindled a diagram of disease that we have already characterised as exclusionary. Contagionism in that sense is a hybrid of science and culture and was not given in the order of things. Second, they remind us to be wary of triumphalism and the over-confidence it suggests. The power of contagionism and germ theory should not blind us to the vital truths that anti- and contingent contagionists as well as modern social medicine and radical public health intellectuals have emphasised. While far from innocent, given the commercial interests that adhered to anti-contagionist arguments, those opposed to contagionist explanations and interventions were developing an evidenced-based science that maintained a more relational, less reductionist and multi-factorial understanding of infectious disease.

Calls to re-energise configurational understandings of infectious disease have more recently come from both scientific and social scientific quarters. For example, the late molecular biologist Joshua Lederberg (2000) urged for a shift away from the 'war metaphor' between microbes and people, advocating a more ecological understanding of microbial biology. It was a call that was followed up in a US National Academy Report in 2006 entitled *Ending the War Metaphor*, in which the authors advocated a dynamic understanding of host–microbe relations (Institute of Medicine (US) Forum on Microbial Threats, 2006; Methot, 2015). The report approvingly cited the work of immunologists Casadevall and Pirofski, who were developing a 'damage-response framework' for understanding virulence and host-parasite relations (Casadevall and Pirofski, 2003; Methot and Alizon, 2014). More recently, Casadevall and Pirofski (2014: 165) have called for a 'ditching' of the term pathogen altogether, noting that disease is but 'one of several possible outcomes of an interaction between a host and a microbe'. As pathogen-centred accounts of disease have obscured greater understanding of the importance of the range of interactions that make disease a possible outcome, they have advocated a search for new diagnostic and management tools that involve pathologists and immunologists working together.

This attention to the host is welcome, even if the biblical connotations of hosting a disease are hard to ignore, but configuration is and should be broader and require not only pathologists and immunologists, but also social scientists. In medical anthropology, for example, there is an insistence on the role played by social conditions in generating new kinds of 'pathogenicity' (Farmer, 1996: 206). For Farmer, pathogenicity it is not just a matter of pathology and immunology, it is also a matter of sociology and geography. Understanding social inequalities, vulnerabilities, differentials in social and ecological resilience and uneven risk geographies should be part of the tool kit. Configurations involve more than the host and microbe, they are linked to production and poverty, and are in that sense political and economic.

In short, in configuration approaches, disease is no longer only a matter of identifying a pathogen or a transmission route. This takes us somewhat beyond the confines of the infected premise or site and requires that we generate an analysis of the disease situation. It also suggests that we require a social and spatial imagination that is able to generate links between sites. This brings us to the consideration of situations as spatially composite. Situations help us to specify the range of actors and actants that make a disease possible. They are more than biological and a matter of configuration rather than simply an infected premise whose borders to the contingent, outside, world have been breached. Situations, as we will argue, are also more than the co-presence of multiple factors. In order to explore this further we now need to discuss a social theory of situations in a little more detail.

Situations and Social Theory

As we have already noted and as our expanded sense of disease configuration suggests, a situation exceeds individual sites. It is a relating together of matters that can involve other sites, institutions, practices, people, animals, materials and processes. A situation, we suggested, demands that we trace and build relations that are relevant to the disease outcome. In order to be able to do this we need to clarify our understanding of the spatial relations that make a situation. We start with a useful account of what are called site ontologies (Schatzki, 2003), before raising some qualifications to this work by introducing post-humanist, multi-site and trans-species approaches to fieldwork. We finish the section with a brief account of the situated practices and empowered situations of feminist philosophers Isabelle Stengers (2010b) and Donna Haraway (1988).

Sites are clearly, as we have already stated, more than individual cases. They are embedded in and relate to other places, processes, practices, issues and materials. In our terms, they are situated. A key issue for thinking situations has been, to put it starkly, how to relate sites to the processes that help to inform and form them without leaving those processes as somehow disembodied or free floating abstract 'structures', wholes, or emergent levels of description (or indeed scales, see Marston et al., 2005; Jones III et al., 2007). It is to avoid, in other words, society becoming a general category that somehow formats life, individuals or places without itself being located, grounded or a matter that is also in formation. It is also to avoid the equally problematic tendency to reify local conditions or context as somehow mediating global conditions. In these accounts the localness of a location is somehow disembedded from the spatial processes that presumably contribute to that location's specificities. The problem is of course one with which many social scientists have grappled and is a reason that we are attracted to approaches like actor network theory which assumes that actors and sites are recursively defined by their network, while networks are defined by their actors or sites (Law and Hassard, 1999). In this approach, actors and networks are on the same level, are accessible to one another, and mark different 'sites', temporary stopping points, or 'attributions' that require us to relate them back or circulate further in order that we do not mistake them for *the* site, or *the* social and so on.

A useful expansion on this more recursive style of reasoning is provided by Theodore Schatzki (2003) in *The Site of the Social*, an attempt to develop a more situated account of social formations through a 'site ontology'. The site ontology draws on Heidegger and Foucault among others to insist on the primacy of embedded practices as constitutive of social order. It thus refuses to see context or the social as somehow divorced from or separate to grounded practices. The continuously practised or enacted nature of the site of the social seeds an understanding of a changeable, process-based, embedded and embodied social life. The site of the social is hence a 'continuously churning enveloping horizon of organised human activity' (Schatzki, 2003: xii). It is in this sense 'a thoroughly contingent and a severely fragmented affair' (xii). But, to be clear, it is far from fractured and individual, for the site of the social persists through an albeit contingent 'mesh of orders and practices' (xi). The social is a 'site-context' in which *orders*, or what Schatzki sees as arrangements of entities (e.g. people, artefacts, things), are combined with *practices*, a term he reserves for organised activities.

Schatzki's account usefully embeds social activity within sites in a manner that is recursive. There are however several issues upon which we might want to reserve judgment in a study of infectious diseases and social situations. First, in Schatzki's account, the distinction between orders (arrangements of entities) and practices (organised activities) is more than a heuristic. He associates orders with arrangement or network theorists (including ANT) and those who elevate assemblage, apparatus and/or agencement as the key means through which the social is performed (e.g. Latour, 2005b). Foucault, Latour, Callon, Deleuze, Guattari, all, according to Schatzki:

> ...highlight the elementariness of what might be called the labyrinth 'configurational order' of the social: the involuted lacing of human and other phenomena into extensive arrangements that determine as well as bind together their characters and fates (Schatzki, 2003: xiii).

For Schatzki, who by no means denies the powers of arrangements and confirms the 'existence of non-human agency' (2003: 71), these arrangements are nevertheless largely derivative. For 'practices establish particular arrangements. These arrangements are definite packages of entities, relations, meanings, and positions, whose integrity derives from the organisation of practices' (2003: 101). Moreover, practices are conceived of as an organised nexus of activities of 'doings and sayings' (2003: 73) and are, for Schatzki, *strictly human* affairs. In this he explicitly distances himself from those in science studies who suggest practice is something that results as much from non-human as from human agency (Pickering, 1995). For Schatzki, this is 'a point of determination' rather than a moral claim (Schatzki, 2003: 71). Contrary to Pickering, and for Schatzki, there is no *mangle of practice*, or the production of difference through the shared imperfections of human and non-human actions, ordered or otherwise. The 'point of determination' is that human beings' doings and sayings are the key drivers of social life.

In privileging practice (organised activities) over arrangements, and humans over non-humans, we are left with a comforting if problematic account of the site of the social. Humans, and one presumes some humans more than others, seem equipped to practise and innovate while most of the rest of the world needs to be content with arrangements. Leaving aside questions of who gets to act in this account, empirically, we would argue, we can imagine cases where practices, human or otherwise, are far from determinant in terms of how things take shape. Perhaps the problem here is with the word 'determine'. We do not read arrangements as 'determining' character or fates – indeed this would contravene one of the first rules of most post-structuralist inspired versions of STS regarding the inappropriateness of technological or other forms of determinism. Likewise, while 'the specific character of social life is to a remarkable extent *attributable* to these bundled activities [human practices]' (Schatzki, 2003: 71 emphasis added) that is not equivalent to suggesting that such activities are determinant or causal, or indeed primary in anyway. We would prefer to see practices and orders (or orderings following Law (1994)) as embedded within one another rather than as discrete divisions between, on the one hand, process and on the other, form. Humans, whether privileged or not, are after all always constituted through their relations, and these relations are always more than human. Our point then is to remain empirically agnostic on the issue of agency and social life. Situations, for us, will be more than human both in terms of their fabric and in terms of their determinants.

A second issue is somewhat less problematic, and surrounds an omission rather than a point of departure. Schatzki rightly highlights the importance of empirical work to social theory, following Barnes's insistence that 'theory without some kind of exemplification is no theory at all' (Barnes, 1995: 61). But the exemplifications that Schatzki chooses seem rooted rather than routed, or located at single sites rather than distributed across numerous settings. In a globalising world, place-based work should provide a means to evoke pathways to other places (Marcus, 1995). Anna Tsing (2004) calls this the challenge of doing an ethnography of global connections. Ethnography was of course originally designed for what were conceived of as small and relatively tightly bounded communities, so its adaptation for global connections requires something more 'haphazard', a travelling theory that is alert to 'the zones of awkward engagement... zones of cultural friction' which are transient and 'reappear in new places with changing events' (Tsing, 2004: xi). Tsing's concept of friction draws attention to the importance of contact as the means through which social change is effected. But this is not the contact of contagion or contamination, it is rather a relational matter wherein the qualities of the interfaces affect change. Rather like a spinning wheel which goes nowhere prior to contact with another surface, ideas, theories, ways of doing things, can only travel, can only exert influence, through contact. Friction is key to things becoming effective, but it also has a price as friction is not only about traction, it is also about awkward exchanges, a rubbing together of differences. It is those interfaces that make globalisation, modernisation, security and so on both possible and always in process, so much so that they are refigured as interchanges rather than diffusions of unchanging universals. These contact points may produce novelty, newly charged situations or even new possibilities for organisms, organisations and so on. This is a matter, then, of contingency, the touching together of differences, which often produces surprising outcomes. In short, frictions can be and are constitutive of the new, and give rise to contingency. They form the privileged sites, in Tsing's account, for a multi-site ethnography.

Finally, combining these issues together we are struck by not only the need to evoke or abduct from the multi-sites of a situation (the corporate offices as well as the farms, the breeding grounds of birds as well as virological laboratories in our case of the Driffield ducks), we are also intrigued by the multi-species or better trans-species nature of avian influenza and other disease situations. As Kirksey and Helmreich put it: 'With animals, invasive plants, and microbes on the move, anthropological accounts ramify across places and spaces, entangling bodies, polities, and ecologies' (Kirksey and Helmreich, 2010: 555–556). As we have seen in the Driffield duck example, to understand the avian influenza situation we need to travel, and look for sites of friction, but we also need to range across 'species'. Wild birds, ducks, chickens, people, viruses, proteins, antibodies, margins, diets, trade – all are entangled in the situation. Their intra-actions and dynamics mean that even talking of species seems rather redundant (Dupré, 2012; Landecker, 2015). Even so, contra Schatzki, these assemblages are more than arrangements or orders. We take these trans-species matters, or quasi-entities, as awkward entanglings that can cause friction and re-animate a politics of life. We do not see them as secondary to human 'doings and sayings' but part and parcel of a broader cloth of chattering matters. As we will see in the next section, viruses, hosts and infrastructures might be said to chatter too. It is not only humans who 'do' or even who have a say.

Before we turn to these doings and sayings of the more than human, multi-sited and trans- or post-species situations, we want to add one more facet to our use of the term disease situation. For we take the term to be not only one that traces an ontology of the social, it also invokes for us a politics of fieldwork and of academic practice. Here the work of feminist scholars has been formative in moving academic work from what might be termed the (somewhat passive) following of relations to the identification of key moments of friction (Tsing, 2004), the identification of situated knowledges (Haraway, 1988), and the requirement to reason back from those sites and moments in ways that generate a situation, or, in the terms of Stengers (2010b), empower a situation. We have already touched on Tsing's image of friction and the resulting identification of key sites in global anthropology. Haraway's employment of 'situations', as a means to refuse relativist as well as unconditional or ahistorical knowledge claims, helps to underline the importance of an abductive logic through partial sharing of faithful accounts. As she states:

> I think my problem, and 'our' problem, is how to have simultaneously an account of radical historical contingency for all knowledge claims and knowing subjects, a critical practice for recognising our own 'semiotic technologies' for making meaning, and a no-nonsense commitment to faithful accounts of a 'real' world, one that can be partially shared and that is friendly to earth-wide projects of finite freedom, adequate material abundance, modest meaning in suffering, and limited silliness (Haraway, 1988: 579).

'Situatedness', for Haraway, is a means to account for the powers of doings and sayings at the same time as acknowledging the ways in which the trickster-, coyote- and Frankenstein-like objects of knowledge will transform those doings and sayings and play a part in knowledge practices. But the phrase we would highlight from the extract above is 'commitment to faithful accounts'. There is, in short, a normative element to situated knowledge for Haraway, one that suggests that anything does *not* go, and that it is incumbent on science, including social science, to be responsive to their situation.

This normative impulse requires that we not only trace the situations and situatedness of a disease, we also seek to stay faithful to that situation. Doing so requires, in Stengers's terms, empowering a situation in ways that are attendant to those awkward entanglements of people and non-humans that force them to think (Stengers, 2010b). In other words, it is the humans *and* non-humans who are bound together in practices that can generate a situation. Moreover, it is one role of social science to make sure that those imbrications are not lost from view in the normal, or majoritarian version of business and politics as usual. In Stengers's sense we should allow the obligations that people have to non-humans (the knotted ties that have evolved out of shared practices, be they with wild birds, viruses, domestic animals, profit margins or other 'species' of non-human) to enter the political fray. In doing so, she insists that we remain faithful to the events of those obligatory ties, the ways in which they have the power to force those gathered around them 'to think and invent' (Stengers, 2010b: 21). We need to construct what she calls an ecology of practices, an ecology that allows those relations to place established concepts and knowledge at risk. Such an ecology of practices leads not to the empowerment of stakeholders, but of situations, 'giving a situation that gathers the power to force those who are gathered to think and invent' (Stengers, 2010b: 21). Situations are then more than descriptions of affairs, they involve taking the awkward interfaces and contacts between sites, people and non-humans and linking them together, giving them due attention and generating new insights.

Empowering a situation involves taking sites or individual controversies, particular formats of an interface between actants (for example, viruses and birds within a laboratory or in the field) and relating them to an evolving political situation. This relating together is not necessarily focused on their specific details, but on their power to open up or suggest questions surrounding more general conditions of operation (Barry, 2013: 81). By employing this abductive logic and linking sites to situations, we can remain faithful to the event of the site and start to generate disease situations in ways that may challenge or reconfigure the diagrams of disease that we traced in Chapter 2.

We have now expanded on what we mean by disease situations, emphasising that they involve, first, a shift from contamination to configurational understandings of disease and, second, a social and spatial imagination that can generate a multi-sited and trans-species account of disease events. We now need to attend to some current understandings of disease and in particular to pathogenicity as a relational achievement in order to allow us to trace some of the key components of disease situations. Our question is what converts microbes into 'pathogens', what gives them pathogenicity and starts to produce, in our terms, a disease situation? Answering this question involves bringing microbiological understandings of the societies of microbes together with a social science understanding of the societies of people, animals and economies.

Microbial Life and Contagion as Difference and Repetition

In Chapter 2 we noted that the term pathogen, which started to be used in the late nineteenth century, referred to a microbe that could cause disease. As we noted, the tendency is to imagine a self-similar package of disease-inducing matter that, if introduced to an organism from outside, will cause sickness and possibly worse. This is a world, as we have noted, of defined objects with pre-established identities that collide in space. Space in this account is rather like the smooth surface of a billiard table, with the microbe balls striking one another and colliding with and then entering the larger organisms before re-emerging, en masse, essentially unchanged. It is a geography of surfaces and collisions. It is a world where barriers seem to be the best means to halt collisions and stop infection – a one-world world where the centralised structures of global agriculture attempt to seal themselves off from the contaminations and contingencies of a world of circulating microbes. Unfortunately, microbial life and infection is rarely so neat and tidy. Rather than this physical world of collisions, infectious disease may have more in common with a chemical world of reactions, compounds and complexity. To give a flavour of what we might usefully call the intra-active nature of pathogens, or better pathogenicity, we will return to the avian influenza viruses that partly animated our Driffield ducks case.

Avian influenza, originally known as fowl pest and first reported in the 1880s, was linked to a virus in 1955 and subsequently classified as a Type A influenza virus (orthomyxovirus). It was then shown to be a closely related to other influenza viruses that commonly infected people, pigs and horses (Swayne and Suarez, 2000). The avian flu virus is classified into one of two broad types based on its pathogenicity, or in this case the ability to produce 'high mortality and necrobiotic, haemorrhagic or inflammatory lesions in multiple visceral organs, the brain and skin' (Swayne and Suarez, 2000: 463). This pathogenicity is a measure that is standard and based on a specific host (chickens or *Gallus gallus*). To be highly pathogenic one or more of the following criteria must be met:

- any influenza virus that is lethal for six, seven or eight of eight (>75%) four- to six-week-old susceptible chickens within ten days following intravenous inoculation with 0.2 ml of a 1:10 dilution of a bacteria-free, infectious allantoic fluid;
- any H5 or H7 virus that does not meet the criteria above, but has an amino acid sequence at the HA cleavage site that is compatible with HPAI viruses;
- any influenza virus that is not an H5 or H7 subtype and that kills one to five of eight inoculated chickens and grows in cell culture in the absence of trypsin (Swayne and Suarez, 2000: 464).

Two issues of note here are, first, that there is a clear and standardised means to differentiate between highly pathogenic avian influenzas (HPAI) and low pathogenic influenzas (LPAI). Second, this distinction is to a large extent based on the observed or imputed behaviour of the virus in a particular 'model' host, the chicken. This classification is based on a relational effect, not it should be noted on an essential characteristic of the virus. The virus is highly pathogenic in relation to chickens in experimental conditions. It does not follow that an HPAI will be dangerous in a duck, a teal or a person. It probably follows that it will be dangerous to other galliformes (chickens, quail and turkeys), though field conditions may alter the severity of the condition. Conversely, an LPAI may prove to be dangerous to hosts that are not chickens (though this has rarely been demonstrated), or indeed to chickens that are not in experimental conditions and who may be suffering intercurrent infections or disease. Moreover, a low pathogenic strain can mutate to a high pathogenic strain, something that occurred in the Italian poultry industry in the late 1990s. So the classification is neither essential nor timeless (this is the purpose of the second criterion), which effectively notes that some H5 and H7 avian influenzas have a potentiality to shift from low to high pathogenicity. It is a system that is clearly based on the requirements of an international, trading poultry industry, and is devised to provide notification of the presence of a dangerous disease to the broiler and chicken egg layer industries and to the governing and trade bodies who sanction 'disease-free' or 'not disease-free' status in national and regional animal populations. It is worth noting that chickens, as well as turkeys, are not thought to be established hosts for avian influenzas, which gives some sense of why infection in these domestic sub-species can result in such intense disease signs (for more detail on influenza and attachment sites see Short et al., 2015).

This species-focused and economically-derived or trade limiting classification of avian influenzas starts to suggest both a situation-dependent system of naming and defining (with all its limitations), but also the possible variety of situations that a virus will find itself in; different host species, in different environments, along with other host–microbe intra-actions and states of health and so on. So, despite their names and the air of solidity or essential character that they seem to confer, it is worth remembering that a virus is always relational, and that 'a microbe cannot cause disease without a host' (Casadevall and Pirofski, 2014: 165). To paraphrase Spinoza, we clearly do not know what a (microbial) body can do, until, that is, we know something more about its situation.

This Spinoza-inspired 'declaration of ignorance' (Deleuze, 1988: 17) is even more apparent once we examine the avian influenza virus in some more detail. For it is not only variable with respect to its host, it is also variable with respect to itself. Like other microbes, and indeed other organisms, it may be more useful to regard the virus as a developmental process. For, as Bapteste and Dupré (2013: 381) note, 'when we use a set of properties to describe the adult state of an organism, perhaps for taxonomic purposes, we are abstracting a particular time slice from this developmental process.' When the temporalities in question are measurable in milliseconds (as they are for microbial and molecular life) then this abstraction of objects or things, as against processes, becomes even more pertinent. In the case of the avian influenza virus-as-process, for example, every cycle of replication produces thousands of mutations with the potential to alter viral character and its pathogenicity. Avian influenzas are RNA (ribonucleic acid) viruses, with a single strand of genetic material (the RNA strand contains the specific sequences of bases that form the genetic code for the virus). Because these viruses have a single strand of RNA, or code, they are highly susceptible to mutations, or copies going 'wrong'. In simple terms there is no proofreading capability that is usually offered by the second strand in double-stranded sequences. In addition, the gene that codes for the haemagglutinin molecule (responsible for 'entry' into a host cell) is particularly prone to mutation, with the result that infectivity and so pathogenicity may be highly alterable. These mutations make the development of vaccines and the generation of host immunity for RNA viruses particularly difficult, hence the need to develop new flu vaccines on an annual basis.

The result of all this mutability is the production of a viral population or swarm within which there will be genetic differences, and differences in key components of the virus if these changes in code result in shifts in molecular make-up. This swarm constitutes what virologists call 'quasi-species' (Eigen and Biebricher, 1988; Holland, 1993; Lauring and Andino, 2010). For Lauring and Andino (2010: 1), a 'quasi-species is a cloud of diverse variants that are genetically linked through mutation, interact cooperatively on a functional level, and collectively contribute to the characteristics of the population.' There is, then, for these biologists, a society of viruses, a sociality that confers a fitness to adapt to new circumstances. Indeed, for quasi-species theorists, the fitness of a virus may be related to the differences in its repetitions rather than to fidelity to a master copy. In language familiar to social scientists, they talk of a network of variants that makes a population viable. One need not necessarily, though, associate this variance with simple fitness for survival or viruses adapting to niches – for virus fitness also depends on more than infection rates – the survival of hosts and reproduction rates matter too. As population biologists demonstrate mathematically, thresholds become important here as the viral and host population may reach specific densities that shift the population dynamics. These dynamic populations are it seems intrinsically unstable, even given minor shifts in starting conditions. For Robert May (1993: 61) 'it is intrinsic in the nature of most host–parasite relations that one is easily swept into this domain of non-steady cyclic or chaotic behaviour.'

Before we relate quasi-species back to pathogenicity and in particular to the role of conditions (the host and environment) we need to add another layer of complexity. As well as the mutations and *drifts* in a population of a microbe, there are also *shifts* in the genetic make-up of microbes that can occur through recombination and re-assortment. The former involves transfers of genetic material within the cloud of viruses, or transfers that involve a host cell's genome. Re-assortment on the other hand involves genetic transfers when a host cell is infected with more than one strain of influenza virus. The latter is facilitated by the segmented nature of influenza A viruses. In theory, if a cell of a host is co-infected with a seasonal human flu virus population (which is infectious but non-deadly) and an avian influenza (which is deadly but not infectious), the viruses may, in the course of replication, swap segments and produce a new strain of virus that is both viable and able to infect people and cause severe, even terminal illness.

This quasi-species and inter-strain variance and variability become significant in complex ways when we add hosts into the matrix of what makes a virus viable or fit. The latter is not simply a matter of replication and survival (which are both dependent on the ability to co-exist with a host), they also relate to 'immune escape, transmissibility, and cellular tropism' (Lauring and Andino, 2010). Immune escape refers to the ability to bypass the host's immune response, a facility that can be conferred on a viral population that demonstrates enough variance at key antigen sites. Transmissibility refers to the ability of a virus to exit infected cells and to be excreted or shed in a matrix that allows the virus to remain viable in passing from host to host. Finally, cellular tropism refers to specificity that many viruses develop for particular cell types. For influenza these tend to be in the gut of wading birds or the upper and lower regions of the respiratory tract for poultry, but can be more system-wide. A virus may be well suited to infecting a range of cell types, across numerous species, or be highly specific. These tropisms may shift as the society of quasi-species infects new hosts, causing a selection pressure towards a new cellular tropism.

The details of this variance are of course complex, but the point for our purposes is that viral character and make-up will depend on numerous intra-actions that change all parties as immune responses and virus tropisms continually inform or conform one another. Most mutations are redundant, most contact with ill-suited viruses will produce evolutionary dead ends, but given the numbers and temporalities involved, these processes that confer difference through repetition produce enough variance that significant viral change is the rule rather than the exception. And this process is multi- or post-species, not only in terms of the relations between host and virus, but also important intermediaries like bio-films, bacteria, other viruses, antibodies and vaccines and so on. As Bapteste and Dupré (2013: 379) note, hereditary is only one component in the evolution of an entity like a virus:

> Other processes, often involving causal interactions between entities from distinct levels of biological organisation, or operating at different time scales, are responsible not only for the destabilisation of pre-existing entities, but also for the emergence and stabilisation of novel entities in the microbial world.

What this process–view of contagion suggests is two things: first a rejection of the physical model of surfaces and collisions, and second a need to attend to the trans-species and multi-situatedness of disease situations. Viruses, or the swarm we call a virus, seem to learn, read or grasp their situation. And in this they are not alone, as immune responses too can be understood as a learning process, one that relies on microbial life in order to develop and adapt. The nineteenth-century French polymath, Gabriel Tarde (1903), would characterise this as a process of imitation – not it should be said as a matter of emulation, but as a process of forming compounds and a result of unstable interferences (Latour and Lepinay, 2009: 53).

For Tarde, to exist is to differ and invention is borne from repetition, attendant mutation and adaptation. Tarde's monadological view of society, which does away with parts and wholes and instead understands societies (of people, or viruses and of multi-species compounds) as processes of continuous adjustment and expansion, involves 'a kind of contamination that moves constantly, from point to point' (Latour and Lepinay, 2009: 9). But contamination and contagion in Tarde are different to the contamination that Rosenberg identifies as a mode of explanation in epidemiology. More than defilement through touching, con-tamination for Tarde is also generative, an opportunity for learning. Moreover, this generativity is conferred not through a single contact point, but the repetitive process of learning through imitation, opposition and adaptation (Tarde, 1903). Tarde highlights 'chatter' as a key process here, referring to the value of rumour and gossip to the economy (Tarde, 1902). But, interestingly, the same metaphor has arisen in virology. Wolfe and colleagues for example suggest that viral chatter, or the process of viruses and hosts repeatedly engaging in the to and fro of molec-ular conversations, learning one another's biology, is vital to the process of viral evolution. High rates of chatter, in conditions that are favourable, may increase viral diversity and probability of transmission of a virus than can replicate (Wolfe, 2011; Wolfe et al., 2005, 2007). Moreover, in terms of potential for zoonoses or the crossing of species boundaries, 'recent work in genetics suggests that zoonosis may only emerge and become established in humans after a process of repeated, unsuccessful transmissions' (Brown and Kelly, 2014). In that sense, microbes do not simply diffuse across space, or extend across a plane, they make and are made in intensive couplings – ones where repetition and difference count as much as touching.

Situations are in this sense vital. It is not only the presence of a virus or other microbe that is key to understanding a disease emergency, it is the conditions for unstable interferences between the network of viruses and the trans-species domain that it helps to constitute through a chattering process. Pathogenicity is, as Farmer reminded us, about living conditions as much as an essential characteristic of a virus or bacteria (Farmer, 1996). For Farmer, interested in diseases like tuberculosis (TB), poverty makes the expression of symptoms and the transmissibility of the *Mycobacterium tuberculosis* more likely. Configuration is key – not only in that ill health is more likely in some circumstances than others once a pathogen has been introduced. There are also effects that a change in configuration can have on the molecular nature of a microbe. The mycobacterium associated with TB can become hyper-virulent under certain environmental conditions. Poverty, it turns out, has molecular consequences. Such intra-actions between a host and microbial population are well known with respect to anti-microbial resistance, where anti-biotic treatments effectively change the selection pressures for a bacterial swarm and can eventually produce a network of drug resistant microbes (Landecker, 2015), but it is equally likely to occur once the conditions of repetition and difference are altered in other, sometimes more subtle, ways. Rapidly altering the numbers and conditions of possible hosts for avian influenza may, for example, increase the probabilities for a shift in the viral network (Wallace, 2009). Disease situations matter.

We have now started to shift attention away from pathogens and towards pathogenicity and have demonstrated that the latter is co-produced in disease situations. The practices and re-arrangements that make pathogenicity are multi-sited and trans-species. We have gestured throughout to a change in the coordinates of epidemiology. From the surfaces and collisions of billiard tables, pathogenicity has more in common with a chemical world of reactions, compounds and complexity. Our remaining task for this chapter is to characterise the spatiality of this compound world. What, in other words, makes for a useful spatial vocabulary when interrogating disease situations?

A Topological Disease Situation

A disease situation is more than a site, it is also more than a site of the social (Schatzki, 2003) – it is a trans-species, multi-sited matter where practices and orderings are more than human affairs and pathogenicity is produced from an array of seemingly complex interrelations. Pathogens, when seen in this light, do not simply spread out from a mappable location or source, and nor do they merely diffuse across space. They intra-act (to re-use Barad's term (Barad, 2007)) to the extent that they not only mutate as they move, they also drift and shift through all manner of transfections, transductions and transformations. In this version of networking (the active verb has always been important in, for example, actor-network theory (Law, 2004c)) the virulence of a flu, for example, will alter as it, or its quasi-species network, moves.

Significantly, for our argument, such movements are expressed through spatial intension rather than spatial extension; that is, they emerge through the intensity of the relationships (the viral chatter) which compose the spaces of which they are a part, rather than through their extended distribution in a networked space (Allen, 2009, 2016). This is not a world of flat surfaces, with well-defined proximities to sources of infection marked accordingly, but rather a topological landscape of embeddings and disembeddings, where disease registers its presence through intra-actions. Similarly, disease response, state-sponsored or otherwise, is not something we should necessarily imagine in terms of networks extending over space. Good responses may need to be attendant to more than centralisation and control over space (McCloskey et al., 2014). It may need to attend to the interferences and distortions that make a disease situation. Given that these spatial relations are not conducted on a surface, or through a fixed geometry with measurable distances between microbe and hosts, we need to employ a different spatial imagination.

We do this in three ways. First, we mark a heuristic distinction between the geometry of pathogens and the topology of pathogenicity. Second, we outline the connections, continuities and discontinuities that configure a disease situation. Third, we use this topological imagination to highlight the requirement for field-work to range across a variety of materials and senses in order to engage and produce situated knowledges.

More than Geometry

As we have intimated, a self-similar pathogen moving across a surface space is one that is met with an armoury of barriers, borders and surveillance. But a swarm of mutable microbes and more importantly an object called pathogenicity requires a rather different spatial imagination. We can start to evoke this by mark-ing a distinction (for the moment at least) between a topographical or Euclidean, geometrical understanding of space, and a topological spatial imagination that attends to the multiple relations that produce space. If the former is about a fixed grid in which distance and proximity can be measured, and where health and disease are treated as separable, the latter unsettles these coordinates of and responses to movement. Pathogenicity does not move across space, it is a spatial configuration, a spatial making.

This topological sensibility prompts us to do two things: first, to think about proximity, presence and distance in ways that disturb our sense of what is near and what is far. For this we owe much to Deleuze (1993, 1995), but perhaps especially to a conversation between Serres and Latour (1995) and the analogy of the crumpled surfaces of a handkerchief. Where the flat, well-ironed surfaces of a handkerchief stand in for a geometry of fixed distances and defined borders, the fabric, once folded, draws together threads previously held apart and vice versa. In a topological vein, weaves of cloth that were once close are now distant and, conversely, points previously at separate ends of the handkerchief are now in contact with one another. Viewed topographically, we could take the Driffield ducks, for example, to be spatially distant from other infected premises or indeed avian species in the Netherlands, Germany and of course Korea and Japan. Here the question of infection is largely one of transmission of a virus, H5N8, over expanses of space and across species barriers. For the epidemiologist concerned with biosecurity, the existence of good barriers at the site (boot washes, well maintained buildings that stop the incursion of wild birds and so on) only increase the distance between infected premises. If we were to stand in Driffield with different lenses, ones that were more attuned to the production of a disease situation where the spatial configuration of pathogenicity is the matter for attention, then we might notice the continuities between infected premises, their sharing of practices, of the arrangements involved in moving stock through a business and so on. We might notice the distortions and deformations that occur in the viral quasi-species as it encounters densely packed and rapidly growing bodies. In this rudimentary sense, we have started to fold the epidemiological map which measures extent of spread across space, and started to express different possible connections that are not only concerned with pathways from a to b, but also what makes a and b similarly susceptible and part and parcel of a disease situation.

In some hands, this crumpling of topographical space can be read as a offering a distinction between form and process, apparent coherence and actual complexity (Murdoch, 2005). It can, as Martin and Secor (2014) remind us, be read too quickly as a distinction between fixity and flux, between being and becoming – a world of appearances and actual connections. But this would be too simplistic and underplay a second topological twist.

Connections, Continuity and Discontinuity

If a topographical imagination invites us to track objects across space (a tracking that assumes that both object and space remain largely unchanged by movement), then a topological approach sensitises us to the continual transformations *as well as* the durabilities that make a disease situation. Indeed, it is the simultaneous production of change and stability, or what Adam (2004) refers to as the condition that exists between continuity and disruption, that attracts us to topology and to an application of this approach to disease situations.

For example, an RNA virus (which we now understand as a swarm or quasi-species network) is always different from itself (to exist is to differ) and yet coherent enough to reproduce and function. These viruses seem ideal candidates in this sense for what Lury and colleagues refer to as a topological characteristic of harbouring both 'invariance and intrinsic change (understood as deformation)' (Lury et al., 2012: 8). For they retain integrity despite the many spatial twists and turns undergone. In topological terms, such elements are said to be homeomorphic; they undergo transformations or remouldings wherein connections and disconnections are maintained even when there are shifts in form. If these viruses mutate, yet still retain certain qualities despite the tangled arrangements and distortions that they are subjected to, and if this playful quasi-species retains identity without recourse to a master copy or centre (and in fact transforms in relation to its environments), then we can perhaps best under-stand those quasi-species, and pathogenicity, as a topological field. Infectious agents, in that sense, are a set of potential forms actualised only in relation to the environments engaged. They endure, yet change in different ways, depending upon the nature of the intra-actions involved.

This is not simply a social theoretical turn, far from it. Virologists and others working in bioscience speak of topology in two main areas. First there is the topology of the molecular structure of both viruses and host receptors. So the 'globular' head of the haemagglutinin protein in influenza A viruses, for example, is highly varied, and subject to what are called point mutations as a result of changes in the HA gene in the virus. But, despite this variance, topology and so functionality is effectively conserved:

> The head of the H molecule can accommodate a large number of changes without affecting functions like receptor binding, which is hidden in the centre of the three-fold molecule. Because there is a lot of plasticity in this molecule, it can accumulate many changes to avoid the immune response (Webster, 1993: 39).

This invariance with deformation is matched to an extent in host immune sys-tems, which seemingly allow high levels of mutability in immunoglobulin regions of pre B cells, providing an adaptable system for producing antibodies to the mutable viral antigens (Holland, 1993). Likewise, the receptor sites themselves are discussed in topological terms. For example, the binding of haemagglutinin to host cells occurs with sialylated glycans or particular molecules on the surface of cells in the respiratory tract. These cells are different in avian and human respiratory tissues, and so the switch from bird influenza to human flu seems to be dependent on the virus adapting to the new host via a change in its haemag-glutinin (HA) molecule. However, mutation of the HA molecule does not neces-sarily lead to successful transmission. It seems that the human sialylated glycans express a diversity of structures, with linear and branched molecular chains. But more than this, HA makes contact with a region of the glycan that can adopt different topologies – with the human contact point adopting an open umbrella topology that is different to more 'cone-like', avian topology (Chandrasekaran et al., 2008). In short, this shape shift seems to reduce the ability of an avian influenza to bind to human cells despite any necessary changes in the HA mole-cule. This not only prompts questions about non-human animal models for human infectivity (Short et al., 2015), it also demonstrates that presence or absence of viruses, or indeed their apparent distance or proximity to humans (how close they are in metrical distance but also in terms of species suitability) is not the only spatial register that matters. Their conformation, or shape and intra-actions, are vital too. The point for our purposes is that topological twists configure the disease situation all the way down. They are molecular, embedded in tissues, as well as global.

The second field where topology is used is in evolutionary genetics and analysis. Virologists can produce a 'consensus sequence' for a quasi-species (i.e. a statistically significant representation of the genome of a population of viruses that are isolated from a host) and then perform an analysis that links samples to one another as a means to show their relatedness. In the Driffield case, for example, virus isolates were relatively quickly compared with isolates from other infected premises and from wild birds. In other words, the genetic sequences of either the whole viral genome or significant parts of that genome (for example, the important and promiscuous HA gene) are 'extracted' and compared with one another in order to express degrees of similarity. The Driffield consensus sequence was over 99% similar to the sequences found in the Netherlands and Germany, suggesting that it came from a similar source.

However, it is the minor distortions in genotype and phenotypes that may be of more interest. The inherent mutability of viruses allows for inferences to be made on, for example, the routes followed by, and the frictions of, those viruses. One of the infected premises in the Netherlands in 2014, for example, provided a viral signature suggesting that the infection had come from another farm. The virus (or the consensus sequence) had changed enough to infer that it had adapted to life with chickens (European Food Safety Authority, 2014: 2). In other words, the quasi-species shifts that we would expect to occur once 'it' has undergone intra-action with new hosts, are readable in the sequence data of the viruses.

This mapping of genetic similarity and divergence is often used to generate likely lineages and construct phylogenetic trees (see Figure 3.1). The underlying assumption here is that the samples are related through vertical evolution (clonal expansion) that occurs in time – that is, mutations continually drive variation and, in response to changes in selection pressures, eventual stabilisation into new strains or clades. These trees are, though, unstable, and can be changed quite significantly by inputting or deleting a single virus sequence, or indeed changing the means of inferring a tree from the available data (using different statistical techniques and models). They are also dependent on adequate samples, robust sequence data (itself subject to the methods and attendant indeterminacies of sampling a quasi-species population and producing consensus sequences), and attendant metadata (including sampling date, species, location and so on). As most sampling is outbreak dependent, that is, there is little data that refers to the periods of time between outbreaks, or indeed the large number of cases that may go unreported, then data can be sparse. So, for example, the apparent genetic proximities between distant samples of the domestic European and wild bird Japanese H5N8 viruses were difficult to explain. As the resulting official report noted:

> ...reliable interpretation of the topology of the European and Japanese cluster cannot be made with these similar sequences. [Nevertheless, it is hoped that] Phylogenetic analysis of other viral gene segments and importantly sequences from more viruses will help to resolve these relationships (European Food Safety Authority, 2014: 2).

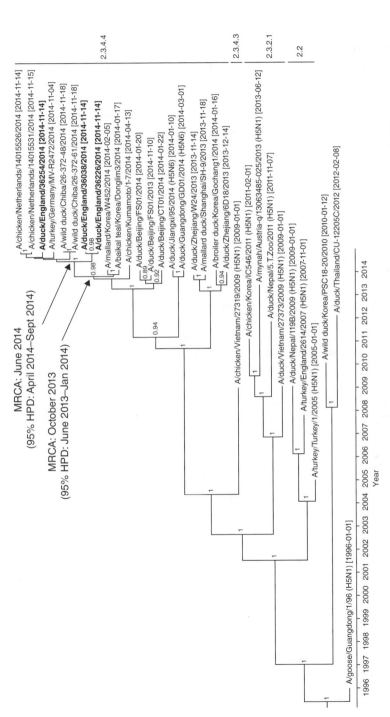

Figure 3.1 Maximum clade credibility tree of 31 H5 sequences derived from the haemagglutinin gene of avian influenza viruses (1,608 nt). Sampling dates and locations are included on the tip labels; where specific dates were unknown, '01' was assigned. Node labels indicate significant posterior probabilities (>0.75). The dates for the most recent common ancestor (MRCA) of the currently circulating viruses circulating in Europe and Japan are indicated at the relevant nodes with 95% highest posterior density (HPD) levels. Sequences relate to H5N8 subtype, unless otherwise noted. The bold sequences relates to the Driffield ducks (Hanna et al., 2015).

More fundamentally, trees tend to impose a version of vertical evolution that may be difficult to justify. They underplay what are called 'reticulate events, such as hybridisation, horizontal gene transfer, recombination, or gene duplication and loss' (Huson and Bryant, 2006: 254). In other words, assumptions of vertical lineage neglect the horizontal intra-actions, or sociality, of microbes. Here, the branching of the phylogenetic tree gives way to network approaches, and to a topology of loops and holes, which may demand different kinds of mathematical analysis (Chan, 2013).

For analytical purposes, and in terms of RNA viruses at least, these networks are often kept simple (and often approximate to trees in terms of their simple branching). For example, Lee et al. (2015) trace the genetic distance relations between H5N8 HA sequences and the H5N2 viruses that severely affected North American domestic poultry in 2015 as a network. While their topology remains tree like, the intercontinental groupings of viruses demonstrate genetic proximity for otherwise geographically distant samples. So, a Japanese wild bird virus is clustered alongside the Driffield and other European samples, suggesting genetic proximity is not solely a function of geographical distance.

While there is a tendency within these analyses to revert to a Euclidean geometry and contagionist aetiology, there are nevertheless openings here for a multi-dimensional approach to pathogenicity. In short, the statistical rules of parsimony and credibility should not wholly distract us from the other spatial relations that may apply to a disease situation. A focus on wild birds as vectors for H5N8 and H5N2 should not, it seems to us, give cause to abandon any questioning of the role of commercial poultry or other aspects of this disease situation in the cycling of microbes and the dynamicity of the disease. Pathogenicity and indeed the dynamicity of host–microbe relations are matters to be explained rather than simply mapped.

To some extent this work of linking deformations and distortions of viral quasi-species to the bio-social clusters that help to direct those changes is under way. Hogerwerf et al. (2010) have broadly characterised and mapped the agro-ecologies where there is persistence of highly pathogenic avian influenza. In their spatial and quantitative analysis, key drivers seemed to be chicken and duck populations and densities, as well as purchasing power and the commercialisation of farming. Their conclusion was that *persistence* of HPAI was *not* something associated with poverty (and low biosecurity), but with a mixed landscape in which there was a significant presence of commercial and intensive poultry, (a conclusion that is backed up by Dixon's (2015) detailed mapping of the spatial flows that exist within and between parts of the poultry sector in Egypt).

Our task has been not to replicate such studies, but to generate detailed understanding of the socio-economic and regulatory processes that can produce such persistence and may, as we have suggested, generate further possibilities for disease emergence or pathogenicity. In the final sub-section we summarise some of the methods that we have used as a means to take these disease situations seriously.

Studying Disease Situations Topologically

Drawing these excursions into topology together we might say that the shape and size of things or the distance between them is less significant than the relationships which tie them together (Allen, 2011). To be sure, this re-imagining of pathogenicity and disease situations is not straightforward, but it does involve a rather different approach to fieldwork on the security or otherwise of a disease situation. Again, rather than approaching a disease site with a focus on transmission and barriers, we can use a topological imagination to interrogate a multi-dimensional matrix of matters that make a disease situation. This allows us to work at new ways of situating disease. Shields, for example, links a topological sensibility to this reformatting of diagrams: Topology 'provides the mental hand-holds for working with situations where relationships are changed, distanciated, collapsed or distorted, reshaping the "diagram" one might draw of the situation' (Shields, 2012: 48). It is an opening, then, onto a way of seeing that moves us beyond the site, generating 'novel and non-totalising hypotheses which shed light on the dynamism of experience and perception, norms and practice' (Shields, 2012: 49). For us, it also reminds us that there is more than one diagram and more than one spatiality to any disease situation. Unlike the barriers and centralised systems of McCloskey et al. (2014), which seem to prioritise an exclusion diagram of disease and in doing so promote a topographical and regional spatiality, we are interested in what Mol and Law (1994) characterised as inter-topological effects. For Mol and Law, the world is enacted through a variety of topologies and attending to only one of these can reduce our ability as fieldworkers to see the range of activities and spatialities that make a situation diseased or otherwise.

Researching this dynamism involves, as we have already intimated, being in more than one place (multi-site) and attending to more than one kind of entity (trans-species), but also, now, being alive to the spatialities of disease. The latter are matters of intension as well as extension, and render suspect any simple one-to-one mapping between the presence of pathogens and disease. Pathogenicities are compound matters.

To be clear, this is not simply a matter of adding more variables into the model (on this see Chapter 9). Intra-actions are not about a coming together of already existing entities. They involve diffraction and unstable interferences or 'the mutual constitution of entangled agencies' (Barad, 2007: 33). There is a dynamicity here that is conferred through all manner of relations. In the terms of pathological lives, we need to engage this dynamic that is borne out of 'the ongoing materialising of different space-time topologies' (Barad, 2007: 141).

To investigate these materialising topologies requires an ethnographic approach that involves speaking with others, listening, mapping, staging deliberations as well as experiencing things for ourselves. It has taken us to farms, to laboratories, to government offices, to veterinary practitioners, to official meetings, to wild bird reserves, to conferences, to retailers and processors, to restaurants, to people's homes, but also into the biologies of guts, cells and microbes. In each place we have been attendant to practitioners' worlds, to what forces them to think and to be affected by their non-human companions, be they wild birds, profit margins or whatever. In being there and in formulating rationales for abducting field sites and relations into disease situations we have been keen to generate a feeling for those situations. This is, as Dixon and Jones remind us, more than a revisualisation (Dixon and Jones III, 2014). We agree that thinking topologically and diagrammatically can reproduce a rather too clean and ocular sense of the social. As we will demonstrate in the following chapters, with their chicken guts, pigs' bodies, cuts of meat and senses of emergency, disease situations are multi-sensual. They are 'tactile' as Dixon and Jones suggest, and require us to attend to the 'various materials and forces that grab onto each other, interpenetrating and reassembling at various speeds and intensities, such that diverse and proximate distances, contacts and connections, are made and remade' (Dixon and Jones III, 2014: 1). But this sense of interpenetration may also be overdone. Chattering microbes are more than likely to pass by without being able to grab at a receptive tissue. It may be some time before the exigencies of a profit margin start to reconfigure immune responses and viral intra-action. A disease situation is then, for us, characterised by a thickened space that urban theorist Saskia Sassen suggests is a 'dense and complex borderland marked by the intersection of multiple spatio-temporal [dis]orders' (Sassen, 2006: 392). These are spaces in the making, frontier zones as she calls them, where borders are continually being restated through the juxtaposition of different elements, some close up, others folded in from afar, detached and re-embedded in ways that give rise to new and novel arrangements through different types of engagement. Disease is not Sassen's concern, but her depiction of a borderlands where contrasting elements working to different rhythms and logics come together, or better still, intra-act, captures nicely the topologies and intensities of disease situations (Hinchliffe et al., 2013).

Our role then as ethnographers of disease situations is to abduct sites in ways that are empirically robust and verifiable, and remain faithful to the disease events that we engage. The latter has required that we check our understanding and fold our interpretations back to practitioners, that we build situated knowledges (Haraway 1988), but at the same time maintain a topological sensibility to the generativity of the connections that we make, and so aim to empower the disease situation.

Conclusions

Disease, we have suggested in these opening chapters, is always and already embedded within life. Life is pathological. It is not something that can be excluded or treated as a contaminant or a microbe that is 'out of place'. That disease can often be diagrammed as outside to good life is challenged by what we have referred to here as topological disease situations. Topologies unsettle these coordinates of near and far, or trace them in different ways. Presence and absence of pathogens become displaced by an investigation into the numerous embeddings and disembeddings that increase or attenuate pathogenicities. Living with disease, if that is the right term, requires that we improve our understanding of disease situations. The next five, more empirically-based, chapters focus on particular disease situations each with the aim of helping us to re-imagine how the disease emergency with which we started might be understood. Their particularities will be different to the avian influenza case that has been used as an exemplar in this chapter, though we return to the socio-economy of poultry in Chapters 4 and 8. But the principles of configuration and their molecular as well as socio-economic topologies unite the cases and our efforts to understand and intervene in the dynamicity of pathological lives through disease situations.

References

Ackerknecht, E.H. 1948. Ancontagionism between 1821 and 1867. *Bulletin of Historical Medicine*, 22, 562–593.

Adam, B. 2004. *Time*. Cambridge, Polity Press.

Allen, J. 2009. Three spaces of power: Territory, networks, plus a topological twist in the tale of domination and authority. *Journal of Power*, 2, 197–212.

Allen, J. 2016. *Topologies of Power: Beyond Terrritory and Networks*. London: Routledge.

Animal and Plant Health Agency 2014. *Epidemiology Report: H5N8 Highly Pathogenic Avian Influenza Outbreak in Breeding Ducks*. UK: https://http://www.gov.uk/government/uploads/system/uploads/attachment_data/file/412086/ai-epi-report-nov2014.pdf

Bapteste, E. & Dupré, J. 2013. Towards a processual microbial ontology. *Biol. Philos.*, 28, 379–404.

Barad, K. 2007. *Meeting the Universe Halfway: Quantum Physics and the Entanglement of Matter and Meaning*. Durham, NC: Duke University Press.

Barnes, B. 1995. *The Elements of Social Theory*. Princeton NJ: Princeton University Press.

Barry, A. 2013. *Material Politics: Disputes Along the Pipeline*. Oxford: Wiley Blackwell.

British Poultry Council 2014. *Economic impact assessment: The British poultry industry 2013*. *In*: Economics, O. (ed.), Oxford: http://www.britishpoultry.org.uk/wp-content/uploads/2014/03/Economic-Impact-Assessment-2013.pdf

Brown, H. & Kelly, A.H. 2014. Material proximities and hotspots: Towards an anthropology of viral haemorrhagic fevers. *Medical Anthropology Quarterly*, 28, 280–303, doi: 10.1111/maq.12092.

Casadevall, A. & Pirofski, L.-A. 2003. The damage-response framework of microbial pathogensis. *Nature Review Microbiology*, 1, 17–24.

Casadevall, A. & Pirofski, L.-A. 2014. Ditch the term pathogen. *Nature*, 516, 165–166.

Chan, J. 2013. *Network and Algebraic Topology of Influenza Evolution*. PhD thesis, Columbia University.

Chandrasekaran, A., Srinivasan, A., Raman, R., Viswanathan, K., Raguram, S., Tumpey, T. M., Sasisekharan, V. & Sasisekharan, R. 2008. Glycan topology determines human adaptation of avian H5N1 virus haemagglutinin. *Nature Biotechnology* 26, 107–113.

Craddock, S. & Hinchliffe, S. 2015. One world, one health? Social science engagements with the one health agenda. *Social Science and Medicine*, 129, 1–4.

Deleuze, G. 1988. *Spinoza: Practical Philosophy*. San Fransciso, CA: City Lights Books, translated by Robert Hurley.

Deleuze, G. 1993. *The Fold: Liebniz and the Baroque*. New York: Columbia University Press.

Deleuze, G. 1995. *Negotiations: 1972–1990*. New York: Columbia University Press.

Dixon, D.P. & Jones III, J.P. 2014. The tactile topologies of Contagion. *Transactions of the Institute of British Geographers*, 40, 223–234.

Dixon, M.W. 2015. Biosecurity and the multiplication of crises in the Egyptian agri-food industry. *Geoforum*, 61, 90–100.

Dupré, J. 2012. *Processes of Life: Essays in the Philosophy of Biology*. Oxford: Oxford University Press.

Eigen, M. & Biebricher, B.M. 1988. Sequence space and quasispecies distribution. *In*: Domingo, E., Holland, J.J. & Ahlquist, P. (eds), *RNA Genetics*, vol. 3. Boca Raton, FL: CRC Press.

European Food Safety Authority 2014. Highly pathogenic avian influenza A subtype H5N8. *EFSA Journal*, 12, 3941.

Farmer, P. 1996. Social inequalities and emerging infectious diseases. *Emerging Infectious Diseases*, 2, 259–269.

Hanna, A., Banks, J., Marston, D., Ellis, R.J., Brookes, S.M. & Brown, I.H. 2015. Genetic characterization of highly pathogenic avian influenza (H5N8) virus from domestic ducks, England, November 2014. *Emerging Infectious Diseases*, 21, 879–883.

Haraway, D. 1988. Situated knowledges: The science question in feminism and the privilege of partial perspective. *Feminist Studies*, 14, 575–599.

Hinchliffe, S. 2015. More than one world, more than one health: Reconfiguring interspecies health. *Social Science and Medicine*, 129, 28–35.

Hinchliffe, S. & Bingham, N. 2008. Securing life – The emerging practices of biosecurity. *Environment and Planning A*, 40, 1534–1551.

Hinchliffe, S., Allen, J., Lavau, S., Bingham, N. & Carter, S. 2013. Biosecurity and the topologies of infected life: From borderlines to borderlands. *Transactions of the Institute of British Geographers*, 38, 531–543.

Hogerwerf, L., Wallace, R.G., Ottaviani, D., Slingenbergh, J., Prosser, D., Bergmann, L., Gilbert, M. 2010. Persistence of highly pathogenic avian influenza H5N1 virus defined by agro-ecological niche. *Ecohealth*, 7, 213–225.

Holland, J.J. 1993. Rapid evolution of RNA viruses. *In*: Morse, S.S. (ed.), *Emerging Viruses*. Oxford: Oxford University Press.

Huson, D.H. & Bryant, D. 2006. Application of phylogenetic networks in evolutionary studies. *Molecular Biology and Evolution*, 23, 254–267.

Institute of Medicine (US) Forum on Microbial Threats, 2006. *Ending the War Metaphor: The Changing Agenda for Unralvelling the Host-Microbe Relationship*. Washington, DC http://www.ncbi.nlm.nih.gov/books/NBK57071/Accessed 11 August 2015, National Academies Press.

Jones III, J.P., Woodward, K. & Marston, S.A. 2007. Situating flatness. *Transactions of the Insititue of British Geographers*, 32, 264–276.

Kirksey, S.E. & Helmreich, S. 2010. The emergence of multi-species ethnography. *Cultural Antrhopology*, 25, 545–576.

Landecker, H. 2015. Antibiotic resistance and the biology of history. *Body and Society*, doi: 10.1177/1357034X14561341.

Latour, B. 2005b. *Reassembling the Social: An Introduction to Actor-Network-Theory*. Oxford: Oxford University Press.

Latour, B. & Lepinay, V.A. 2009. *The Science of Passionate Interests: An Introduction to Gabriel Tarde's Economic Anthropology*. Chicago: Prickly Paradigm Press.

Lauring, A.S. & Andino, R. 2010. Quasispecies theory and the behaviour of RNA viruses. *PLoS Pathog* 6, e1001005. doi:10.1371/journal.ppat.1001005.

Law, J. 1994. *Organizing Modernity*. Oxford: Blackwell.

Law, J. 2004c. Mattering, or how might STS contribute? Lancaster, UK: Centre for Science Studies, Lancaster University http://www.comp.lancs.ac.uk/sociology/papers/law-matter-ing.pdf

Law, J. & Hassard, J. (eds) 1999. *Actor Network Theory and After*. Oxford and Keele: Blackwell and Sociological Review.

Lederberg, J. 2000. Infectious history. *Science*, 288, 287–293.

Lee, D.-H., Torchetti, M.K., Winker, K., Ip, H.S., Song, C.-S. & Swayne, D. 2015. Intercontinental spread of Asian-origin H5N8 to North America through Beringia by migratory birds. *Journal of Virology*, doi:10.1128/JVI.00728-15.

Lowe, C. 2010. Viral clouds: Becoming H5N1 in Indonesia. *Cultural Antrhopology*, 25, 625–649.

Lury, C., Parisi, L. & Terranova, T. 2012. Introduction: The becoming topological of culture. *Theory, Culture and Society*, 29, 3–35.

Marcus, G.E. 1995. Ethnography in/of the world system: The emergence of multi-sited ethnography. *Annual Review of Anthropology*, 24, 95–117.

Marston, S.A., Jones III, J.P. & Woodward, K. 2005. Human geography without scale. *Transactions of the Insititue of British Geographers*, 30, 416–432.

Martin, L. & Secor, A.J. 2014. Towards a post-mathematical topology. *Progress in Human Geography*, 38, 420–438.

May, R. 1993. Modelling ecology and evolution. *In*: Morse, S.S. (ed.), *Emerging Viruses*. Oxford: Oxford University Press.

McCloskey, B., Osman, D., Zumla, A. & Heymann, D.L. 2014. Emerging infectious diseases and pandemic potential: Status quo and reducing risk of global spread. *The Lancet: Infectious Diseases*, 14, 1001–1010.

Methot, P.-O. 2015. Science and science policy: Regulating 'select agents' in the age of synthetic biology. *Perspectives on Science*, 23, 280–309.

Methot, P.-O. & Alizon, S. 2014. What is a pathogen? Towards a process view of host–parasite interactions. *Virulence*, 5, 775–785.

Mitropoulos, A. 2012. *Contract & Contagion: From Biopolitics to Oikonomia*. New York: Minor Compositions.

Mol, A. & Law, J. 1994. Regions, networks and fluids: Anaemia and social topology. *Social Studies of Science*, 24, 641–671.

Murdoch, J. 2005. *Poststructural Geography: A Guide to Relational Space*. London: Sage.

Pickering, A. 1995. *The Mangle of Practice: Time, Agency and Science*. Chicago: The University of Chicago Press.

Rosenberg, C.E. 1992. *Explaining Epidemics: And other Studies in the History of Medicine*. Cambridge: Cambridge University Press.

Rosenberg, C.E. 2009. Commentary: Epidemiology in context. *The International Journal of Epidemiology*, 38, 28–30.

Sassen, S. 2006. *Territory, Authority, Rights: From Medieval to Global Assemblages*. Princeton and Oxford: Princeton University Press.

Schatzki, T.R. 2003. *The Site of the Social: A Philosophical Account of the Constitution of Social Life and Change*. Philadelphia: Pennsylvania State University Press.

Serres, M. & Latour, B. 1995. *Conversations on Science, Culture and Time*. Ann Arbour, MI: University of Michigan Press.

Shields, R. 2012. Cultural topology: The seven bridges of Königsburg 1736. *Theory, Culture and Society*, 29, 43–57.

Short, K.R., Mathilde, R., Verhagen, J.H., Van Riel, D., Schrauwen, E.J.A., van den Brand, J.M.A., Manz, B., Bodewes, R., Herfst, Sander. 2015. One health, multiple challenges: The inter-species transmission of influenza A virus. *One Health*, 1, 1–13.

Stengers, I. 2005b. Introductory notes an ecology of practices. *Cultural Studies Review*, 11, 183–196.

Stengers, I. 2010b. Including non-humans in political theory: Opening Pandora's Box? *In*: Braun, B. & Whatmore, S. (eds.), *Political Matter: Technoscience, Democracy, and Public Life*. Minneapolis: University of Minnestoa Press.

Swayne, D. & Suarez, D.L. 2000. Highly pathogenic avian influenza. *Scientific and Technical Review of the Office International des Epizooties*, 19, 463–482, http://www.oie.int/doc/ged/d9311.pdf

Tarde, G. 1902. *Psychologie Economique*. Paris: Felix Alcan.

Tarde, G. 1903. *The Laws of Imitation*. New York: H. Holt and Company.

Tsing, A. 2004. *Friction: An Ethnography of Global Connection*. Princeton, NJ: Princeton University Press.

USDA 2015. *Avian Influenza Disease* [Online]. http://www.aphis.usda.gov/wps/portal/aphis/ourfocus/animalhealth?1dmy&urile=wcm%3apath%3a%2Faphis_content_library%2Fsa_our_focus%2Fsa_animal_health%2Fsa_animal_disease_information%2Fsa_avian_health%2Fct_avian_influenza_disease Accessed 4 November 2015: United States Department of Agriculture.

Wallace, R.G. 2009. Breeding influenza: The political virology of offshore farming. *Antipode*, 41, 916–951.

Webster, R. 1993. Influenza. *In*: Morse, S.S. (ed.), *Emerging Virises*. Oxford: Oxford University Press.

Wolfe, C. 2013. *Before the Law: Humans and other Animals in a Biopolitical Frame*. Chicago and London: The University of Chicago Press.

Wolfe, N. 2011. *The Viral Storm*. London: Penguin.

Wolfe, N., Daszak, P., Kilpatrick, A.M. & Burke, D.S. 2005. Bushmeat hunting, deforestation, and prediction of zoonotic disease emergence. *Emerging Infectious Diseases*, 11, 1822–1827.

Wolfe, N., Dunavan, C.P. & Diamond, J. 2007. Origins of major human infectious diseases. *Nature*, 447, 279–283.

World Health Organisation 2015. Warning signals from the volatile world of influenza viruses. *In*: *Influenza*. Geneva, Switzerland http://www.who.int/influenza/publications/warningsignals201502/en/Geneva, Switzerland: World Health Organisation.

Zinsstag, J., Schelling, E., Waltner-Toews, D. & Tanner, M. 2011. From 'one medicine' to 'one health' and systemic approaches to health and well-being. *Preventive Veterinary Medicine*, 101, 148–156.

Part II
Disease Situations

Introduction

The inter-relations between the various elements of contemporary biosecurity – security, liberalism, mutable microbes, and pathogen and anticipatory logics – can generate some rather uncertain effects. Indeed, when these are combined together with a range of issues from market pressures to viral difference, they may make matters less rather than more safe. The result of this heady brew is what we are calling a disease situation.

In Part II we start to utilise this thinking in order to develop a sequence of arguments that, first, **diagnose** the current infectious disease situation and second look for **alternatives** within that situation. We start in Chapter 4 with a particular food-borne disease that has recently emerged within the poultry industry, and which we link to the intensities and pressures involved in a sector that not only has to produce a surplus but do so under particular time constraints. The result is what we are calling 'just-in-time disease'. The rise of disease within the so-called bio-secure farming sector suggests to us the limits of human power over life and allows us to re-introduce elements of Esposito's (2008) thinking to guide our analysis.

We then move in Chapter 5 to a similar set of issues as they apply to the pig industry. While the ecologies and economies of production and pathogenicity are somewhat similar, here we engage with the critical role of a variety of practices in producing safe life. We argue that there are resources here that can aid a rethinking

Pathological Lives: Disease, Space and Biopolitics, First Edition. Steve Hinchliffe, Nick Bingham, John Allen and Simon Carter.
© 2017 John Wiley & Sons, Ltd. Published 2017 by John Wiley & Sons, Ltd.

of the disease situation. This takes us beyond diagnosing pathological lives to an outline of alternatives. The ways in which some farmers and veterinarians attend to their animals and their environments suggest a more complex, immunological approach to safe life. Again we draw on Esposito as well as the work of Law (2008) to help us to think again about how healthy lives are a matter of a patchwork of practices and materials.

This patching together is a theme that recurs in Chapter 6. Here we follow the activities and issues involved in making food safe, and note the vital role of food industry experts and inspectors as they engage with various materials in the food chain. Tending to the tensions of what it is that makes food safe, good, tasty and so on turns out to be a complex material performance. What becomes ever more clear here is that the practices that are essential to safe food (and animals) are increasingly being undermined in a world of highly capitalised but often de-skilled food and health landscapes (some of it undertaken in the name of biosecurity). Here we get a hint of the importance of those who know most about the complexities and nuances of human, animal and environmental health and its multi-dimensional characteristics. There is, in short, craftwork that emerges from these studies that points to an alternative framing of biosecurity.

If Chapters 5 and 6 raise the critical role of key intermediaries in pathological lives, in Chapter 7 we ask a slightly different question. How do people who are not directly involved in the production and monitoring processes deal with the threats of food-borne and infectious diseases? Here we employ a different methodological approach, utilising staged deliberative fora in order to generate data on how people respond to disease situations. A key issue here is the surfeit of information that people are 'bombarded' with regarding food and other health scares. We argue that, in light of this, and as a means to change the disease situation, there is a need to place less policy emphasis on providing more information to 'consumers', in order that they can be more vigilant and alter their behaviour or make informed choices. Rather, we may need to focus on what we call disease publics. The latter are not a collection of responsibilised individuals, but collectives that can address a public health issue through an assembly of methods and approaches across the production chain.

Finally, these collectives and publics are more than human matters. Indeed, they are composed of viruses, wild and domestic animals, environments and much more besides. As a result, we argue that attending to safe life requires a form of sensing that is distributed, that is itself more than human. In Chapter 8 we trace what this involves by looking at the practices involved in 'knowing' birds and viruses as the highly pathogenic avian influenza disease situation developed. Our argument is that safe life requires much more than policing or surveillance of life (or generating more information). It requires a different form of knowing, one that Serres (2008) links to the observational sciences. We use this chapter to develop our distinction between a politics of life (biopolitics), in which norms are policed, and a lively politics (cosmopolitics), in which counter-norms and different modes of existence offer us glimpses of alternative ways of making life safe.

We amplify this distinction in Chapter 9, where we argue for a spatialised and more than human approach to both the diagnosis of pathological lives and the empowerment or actualisation of the situations that we have traced in the preceding pages.

References

Esposito, R. 2008. *Bios: Biopolitics and Philosophy*. Minneapolis: University of Minnesota Press.

Law, J. 2008. Actor-network theory and material semiotics. *In*: Turner, B.S. (ed.), *The New Blackwell Companion to Social Theory*. Oxford: Blackwell.

Serres, M. 2008. *The Five Senses: A Philosophy of Mingled Bodies*. London: Continuum.

under laboratory conditions and the poultry industry has become increasingly frustrated by the difficulty of pinpointing the source of infections. Perhaps for that reason, public attention has focused on meat contamination on site and hygiene corners cut in slaughter houses and food-processing plants; that is, on accidents in what is taken to be an otherwise sound system. But such largely isolated failings have drawn attention away from what is arguably a more intractable problem with factory farmed chicken: churning out a highly perishable product on an industrial scale provides the prime conditions for *Campylobacter* to persist in ways that may be pathogenic to human consumers of chicken meat. In short, it may well be the very way that chickens are produced for slaughter, especially following the just-in-time governing principle, that gives rise to a *Campylobacter* situation, not the all too predictable incidents of carcass contamination on site that happens on a routine basis and can be relatively easy to prevent.

The industrialisation of poultry production, with its intensive, tightly controlled food operation, is on the one hand a symptom of a will to control and exclude 'pathogens'. And yet it may well be the source of pathogenicity. The combined pressures transmitted through the production system, from the supermarket shelf and commercial price points through to the hatcheries and standardised uniformities, ensure that farmers (and chickens) are under constant pressure to match the throughput of numbers on a just-in-time basis. When tens of thousands of birds are packed in industrial sheds, their accelerated lives may leave them with compromised and stressed bodies that increase their vulnerability to *Campylobacter*, and indeed produce the intra-actions necessary to generate new kinds of virulence. It is, in the words of Charles Perrow (1999), an 'accident waiting to happen'– a 'normal accident' in his thinking – the likelihood of which is increased given the way that animal bodies, commercial pressures and microbes become entangled in what amounts to a disease situation. In short, these bodies and processes are pressured in ways that can change their dynamicity and produce a new disease situation.

In this chapter, then, *Campylobacter* and factory farmed chicken form a disease situation, with the poultry industry as a site of emergence. The *Campylobacter* situation has less to do with contamination from an unhealthy outside and rather more to do with the just-in-time pressures folded into an already entangled mix of intra-actions. In the terms we used in Chapter 3, it is configurational. We first set out the range of just-in-time pressures, from commercial through to regulatory and biological, which intensify the production of factory farmed poultry and make *Campylobacter* a possible outcome of the pathological configuration produced. After that, we argue that a *Campylobacter* situation is best understood as a relational effect rather than determined outcome of the poultry production system, one best thought about as a topological landscape of pressure points, not breach points, in which disease can erupt as quickly as it can dissipate. Finally, we contend that in taking control of the farmed bird population and reducing it to 'mere' life (Honig, 2009), the powers *of* life may actually be turned against itself. The outcome is closer to Roberto Esposito's (2008) observation that attempts to secure life may actually end up negating it; that is, running the risk of producing ever greater harms.

This chapter is grounded in fieldwork conducted across the UK supply chain for 'table chicken', and in allied food safety and animal health services. Semi-structured interviews were conducted with poultry farmers, poultry catchers, contract labour agencies, quality assurance and welfare managers, factory floor staff, retailers, microbiologists, vets, food inspectors, government policy-makers, and industry advisors. Three of the five major poultry processors in the UK formed part of the investigation, involving a combination of work shadowing and participation observation on farms and in processing plants. This enabled us to observe first-hand the biosecurity practices, people, materials, animal bodies and infrastructure that intersect in efforts to manage food-borne disease.

Factory-Farmed Chicken and Food-borne Disease

With the industrialisation of poultry production, large-scale producers now control more or less every stage of the supply chain, from breeding and hatching, to growing, to slaughter and processing. In the US, the big feed supply companies dominate, integrating forward up the chain, whereas in the UK, the big food retailers, the supermarkets, control supply lines through their market power, with the major food processors picking up the task of backwards integration (Boyd and Watts, 1997; Godley and Williams, 2008).

Regardless of such differences, however, size speaks volumes when it comes to realising the economies of scale that come through concentration, high throughput and product uniformity. Profitability rests upon low unit costs and turning volume; that is, churning the numbers out cheaply. In the UK, almost 1.3 million tonnes of broiler meat were produced in 2011, or 855 million chickens slaughtered (Crane et al., 2012). In 2012, 145 million broilers were in production at any one point in time, at just 2,350 holdings. The total number of birds placed at those holdings over the year was close to 1 billion (918 Million) (all figures derived from agricultural censuses in both UK and Northern Ireland) (Irvine, 2015). In 2014–2015, month by month production of poultry meat was rising at roughly 2% per annum, with over 70 million broilers being slaughtered per month, and over 100,000 tonnes carcase weight produced per month (DEFRA, 2015). The profits of this high volume if low margin industry are significant. The value of the poultry industry to the national economy is estimated to be at least £2.4 billion (DEFRA et al., 2010; Irvine, 2015).

This UK poultry production is a highly concentrated affair, consolidated into fewer and larger vertically integrated companies that, between them, control almost the entire market for chickens in the UK (Crane et al., 2012; Yakovleva and Flynn, 2004). These large corporate processors are now collectively responsible for over three-quarters of the UK slaughter throughput supplied to the commercial retail market (Food Standards Agency, 2010a). The numbers matter, but our interest lies with the range of just-in-time pressures that they bring with them and which do not always sit well together.

Just-in-Time Poultry

With profit dependent upon churning out the numbers of a highly perishable product on an industrial scale, guaranteed throughput and turnover have become essential. A type of just-in-time system, as outlined by Boyd and Watts (1997), is practised to integrate the rapid throughput of tens of thousands of animals from breeding through to growing and processing, and onto the supermarket shelf. Boyd and Watts (1997) set out how, over a relatively short timeframe in post-war US, the poultry industry changed from one comprised of small independent firms operating on a largely informal market basis to a vertically integrated industry managed through a system of formalised contracts. In place of specialised independents at different stages of the supply chain, large corporates, such as Tyson Foods, integrated the whole operation on an industrial basis under a single ownership and management structure. Tight control over throughput at each stage of the process – from delivering the right number of day-old chicks to contract farmers, to ensuring the collection of full-grown birds to keep up with the factory production line, for example – transformed the industry into a system that they tellingly refer to as 'agro-industrial just-in-time'.

The system is not just about tight coupling and numbers though. A constant throughput of birds is critical to the process, but so too is product uniformity and growth rates, in terms of size, weight and body shape, responding to retailer demands for standardised products of particular quality. As Boyd and Watts (1997) outline, these big corporate integrators have relied upon a whole host of technologies to create a uniform gene pool selected for fast growth and higher meat yields, which together with rigid feed and medicinal regimes, produces birds of the requisite market weight and growth pattern. Advances in animal genetics, the rolling out of protein-enriched diets, and the exploitation of pharmaceutical technologies to improve disease resistance, as well as increased growth rates and slaughter weight, effectively combined to take out the vagaries of biological life that, in past times, disrupted supply chains and prompted swings in profitability across the industry. Under corporate agri-business, over the past 50 years or so, the life times of the chicken population in the US has halved and its growth rate doubled, so that the numbers ballooning through the system just-in-time are also, and this is a key point for our account, at just the right weight and price point.

In the UK, over much the same period, a similar pattern of integration, consolidation and standardisation has taken place in the poultry sector, although in this case driven not by corporate 'integrators' but by corporate retail, in the shape of the large supermarkets (Godley and Williams, 2008; Morgan et al., 2008; Wrigley, 2002). Commercial rather than industrial capital took the lead, with Sainsbury's prominent in pushing the leading food processors to exploit the potential economies of scale available by integrating backwards along the supply chain. Many of the innovations in animal genetics, feed regimes and disease control already referred to took place in the UK at the behest of the supermarkets (Godley and Williams, 2008, 2009). Market power is as a result concentrated in the hands of a small number of big retailers, who exercise arm's length control of the food supply chain, through management rather than ownership-style integration, to ensure that chickens bulk up at just the right time in the right numbers.

There are just five major processing companies in the UK poultry sector, each with contracts to a number of the largest supermarkets for a fairly regular throughput of standardised birds at predetermined prices. The UK-based 2Sisters Food Group is one of the leading processors, together with the US conglomerate Cargill, the Dutch-based Vion, Moy Park, a subsidiary of the Brazilian multinational, Marfrig, and Faccenda, a smaller, privately owned UK business. Each, in its own way, exercises close control over their growing programmes, through a combination of own farms and contract farms, as well as through the ownership of hatcheries and feed mills. As one agricultural manager for a major UK processor recounted:

> We are very much a vertically integrated agricultural system, but the vast majority of the (farm) growers are contract growers… But [the birds] are grown under our management system. We supply the feed, we supply the chicks so we've got our own hatchery, we've got our own parent and grandparent stock, and our own feed mill. So we are very much in control of the growing programmes and the agricultural side. (Interview)

But what the processors control, in terms of numbers and throughput, is not totally down to them; that specification is laid down for them by their customers, the supermarkets. Chicks may fail to hatch, flocks may grow more slowly, or there may be losses to injury or ill health, but the supply to and from the factory has to be kept constant to keep up with customer demand. So at the slaughter and processing plant, despite the potential vagaries upstream, throughput has to be predictable and on time:

> To have 1.3 million birds in here every week, it has to be carefully calculated from a hatching perspective in the hatchery, how many birds are put down on the farms, which farms, just to ensure a smooth supply of poultry into the plant because our customer demands are such we couldn't have a million one week and then 1.6 the next week. It's got to be 1.3 every week. (Interview with major poultry processor)

The logistics of the planning operation, to grow birds to market weight and other specifications on time, involves not only decisions about turnover, but also which type of birds to place and where, how to maximise production per square metre on farms, managing the growth cycle of the birds, scheduling flocks for slaughter, and the line speed adopted in the processing plants. Much of this is calculated backwards, as it were, from the supermarket shelf down to the hatcheries and breeding programmes, so that the processors and, in turn, the farmers are under constant pressure to match the throughput of numbers and live weights on a just-in-time basis. The contractual obligations that pull the whole operation together and serve as a means of control, also bring birds, people and materials into contact under conditions of intense pressure that, in turn, can produce new kinds of disease situation.

Figure 4.1 is a representation of the food chain, and what we want to convey is the extent to which the vertical integration of the chain, often by management if not by ownership, exerts pressure across the piece. As we show, while biosecurity measures aim to turn contacts into regulated exchanges (Chapters 2 and 3),

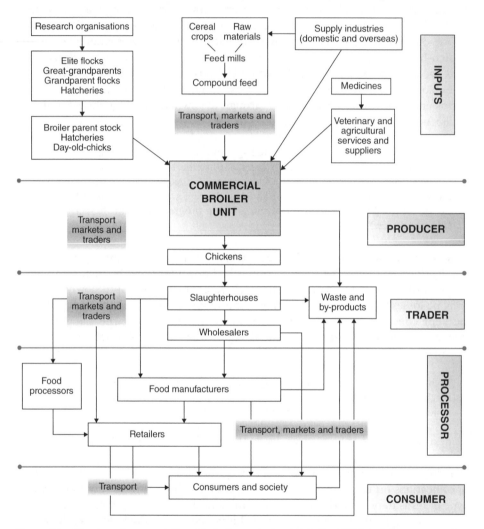

Figure 4.1 This schematic representation of the broiler meat value chain in Great Britain includes a summary view of the component parts of the vertically integrated companies that make up the industry. These broiler integrator companies combine multiple activities from inputs upstream to processing downstream. More specifically this includes feed mills, compound feed, hatcheries, breeder parent flocks, hatching eggs, day-old chick supply, transport, slaughter, further processing and supply of products to wholesale or retail markets. Flow in the reverse direction of the arrows represents the movement of money within the value chain (Irvine, 2015: 146. Reproduced with permission of Elsevier.).

the contracted nature of the businesses acts to constrain or foreshorten the ways in which these exchanges and other living processes can take place.

Securing Sites

The profitability of these integrated systems rests upon delivering safe, disease-free products to consumers, especially given that chicken meat is susceptible to rapid spoiling. The economic efficiencies that size can bring are also often matched by large-scale efficiencies in biosecurity (see Chapter 3 and McCloskey et al., 2014)). After recent scares over avian influenza, as well as greater concern over bacterial food-borne diseases such as *Campylobacter*, the contractual control exercised by the supermarkets has also led to the implementation of tighter biosecurity measures along the food chain (British Retail Consortium and British Poultry Council, 2010; Food Standards Agency, 2010a). The same organisational coupling and integration of the supply chain that is held to be so effective in delivering a consistent throughput of table chicken is also assumed to offer similar advantages of scale and reach when it comes to rolling out hygiene protocols and barrier systems to prevent the incursion of pathogens into live birds on farms and dead birds in factories. This demand for standardisation of biosecurity measures is passed down from supermarket, to processor, to farms:

> Whichever way you do it, [by owning farms or contracting the growing], you make sure you have full control to make changes on farms... Biosecurity is more straightforward with your management structure. (Discussion with major UK poultry processor.)

Barrier systems for changing footwear, boot dips at entries and exits to farms and sheds, hand sanitisation, high pressure washing of vehicles and equipment allowed on site, limiting access onto farms and into sheds: such on-farm measures to prevent the incursion of disease are seemingly as standardised as the factory-farmed birds (Food Standards Agency, 2004, 2006; Red Tractor Farm Assurance, 2010). On industrial poultry farms, biosecurity is thus practised as a bordering process, a disease diagram that focuses on exclusion of pathogens and disease and an attempt to include or incorporate farms and factories within a regulated process. In short, it is a practice of enclosing agro-ecosystems to prevent the introduction of disease.

Retail-led attempts to secure and control, and by this we mean standardise farm and factory practice, is aided by the tight coupling and the contractual integration of suppliers, backed up by the threat of exclusion from the consumer marketplace. Indeed, as we hinted at in Chapter 2, it is the ability to both perform control through a threat of contract withdrawal, at the same time as ensuring brand distance from any system failures that marks out this geography of securing and responsibilisation. Secure and safe produce is a function of a vigilant retail organisation, whose vigilance is performed, they claim, on the behalf of the consumer, though of course this is borne out of a corporate fear of being held to account at a later date (Chapter 2).

Increasingly, retailers attempt to secure market share, and make up for any austerity-led reduction in state regulation or sponsored surveillance of disease, by conducting their own disease audits and accounts of suppliers to ensure that their system is bio-secure. Biosecurity in this sense is a condition of market entry and contract renewal. The growing programmes deliver not only a certain weight and quality of bird at just the right time, but also a flock that is ostensibly safe, 'disease-free', or at least of known disease status with respect to specific pathogens. According to one poultry vet:

> Tesco's probably knows more about the farms than the integrators know about the farms. They will have a file on every farm supplying them, they'll know their mortality, their records, they'll know antibiotics used – because they're absolutely petrified of having anybody saying they've got any problem after eating their food because the financial effect is so great on them. (Interview)

Retailer oversight also includes a regular schedule of site visits to check their standards are adhered to at the factory and farm. As one manager explained as we toured the factory floor, these standards are more onerous even than government requirements:

> The EU only say, 'You have to take care,' whereas the retailers give you a 200 page manual on how to take care. Yes, at least 200 pages. (Discussion with major UK poultry processor.)

This system of checks and routine inspection aims to encourage and perhaps police closure to external disease threats, confirming an order inside the value chain that is able, if it is held to account, to assure the delivery of safe food from farm to fork. Clear reporting lines, where the responsibility for putting their house in order is placed at the farm and factory door, enables the big corporate retailers to shore up their defences against diseases reaching the shop floor, or at least demonstrate that they have taken all reasonable measures to do so.

Scaling-up biosecurity and its surveillance is a modern integrated industry's answer to the incursion of disease, as was noted in previous chapters. Tight integration of the supply chain means that there is no room for operational failure, whether that be a failure of supply of birds or a failure of closure, where an outside world of disease breaks through into 'clean' premises. This 'contamination' approach to understanding and managing disease risk on farms and in factories, based as noted in the previous chapter on the will to remove unregulated contact, is consistent with policy and advice from UK regulatory agencies such as the Department for Environment, Food and Rural Affairs (Defra) and Food Standards Agency (FSA), which understand the task of biosecurity as 'keeping out disease and harmful bacteria' (Food Standards Agency, 2004).

Ready-Made Disease Pools?

The threat of disease, however, is not something that is always seen as external to industrial farming systems. For some, the risk of disease is less to do with the incursion of infectious diseases from the outside and rather more to do with their incubation, amplification and circulation within. An intensive, tightly coupled food operation, on this understanding, may actually harbour disease and accelerate its mutation and spread. In this view, and one which we share up to a point, the disease risk is not so much at our door, as already inside, embedded within modern factory farming (Davis 2005; Leibler et al., 2009; Wallace, 2009).

Life behind the barrier systems in the confined environments of intensive animal production is seen as the problem, in that the dense and standardising ecologies of production are themselves conducive to the circulation, drifts and shifts of microbial lives, amongst increasingly susceptible birds and very possibly people. The sense in which disease may be said to be embedded within modern factory farming thus turns, somewhat paradoxically, on many of the same features already stated. The high throughput of large numbers of animals with short life spans (and rapid growth rates), raised under highly controlled, intensive conditions in confined spaces, rather than offering a more effective, scaled-up biosecure environment is thought to actually pose greater risks for the amplification and spread of viral and bacterial diseases across the many sites of agribusiness (Davis, 2005; Wallace, 2009; Wallace et al., 2009). The routine use of antibiotics within poultry farming, made necessary many would argue on account of the living conditions within the sheds, can itself foment new risks. In 2016, for example, a growth in the use of antibiotics within European poultry, and the switch to specific medicines like fluoroquinolone in the UK, was linked to an increase in *Campylobacter*, *Salmonlella* and *E. Coli* bacteria that were resistant to key antibiotic medicines deemed critically important for the treatment of human infections (European Food Safety Authority and European Centre for Disease Prevention, 2016). Protective barriers to a diseased 'outside', on this view, are of little significance when thousands of confined animals in close proximity seem to provide a ready-made disease incubation chamber. Indeed, a well-honed, biosecure operation may actually make matters worse, not better, simply by 'walling' in disease and intensifying pathogenesis (Hinchliffe et al., 2013).

House tens of thousands of virtually genetically identical animals indoors in densely crowded conditions, pump them full of enriched feeds and veterinary medicines that stimulate growth and weaken their immune systems, and you have what is said to be more or less a perfect ecology for disease incubation (Graham et al., 2008; Lawrence, 2008; Wallace, 2009). It comes down, so the argument runs, to the way that food is produced. The tightly integrated, industrial-scale operations may well provide ample opportunities for contact, transmission and contagion, but just as importantly they may create opportunities for microbes to colonise new spaces, to alter their relation to host cells and tissues and to evolve. Sheer number, uniformity, density and compromised immune systems may, in short, generate different kinds of dynamics for the host–microbe relation.

For Davis (2005), as much as for Wallace (2009), such concentrated animal feeding operations, CAFOs as they are known, are more or less disease cauldrons. The recent spread of the US model of integration, especially across South East Asia and China, is thought to be the reason why disease outbreaks have been so common at large-scale poultry operations in places like Thailand and South Korea, and more recently explains the tendency for H5N2 strains of highly pathogenic avian influenza to occur at commercial turkey, chicken layer and broiler farms in the American Midwest. When tens of thousands of susceptible animals, anything from 15,000 to 70,000 in the case of broiler chickens per shed, are placed at one day old on farms that may comprise of multiple sheds, their accelerated lives may leave them with compromised and stressed bodies, increasing their vulnerability to circulating diseases. As one poultry vet observed:

> The modern bird is very close to diarrhoea shall we say. You're putting a high nutritional value product in one end and you can tend to get looser droppings out of the other. You're growing a 3.5 kilo bird in 38/39 days, which used to take, even ten years ago, would have been five days longer. (Interview)

The mass demand for a cost-effective animal product thus goes further than simply a factory farm model designed for scale and volume, and foregrounds the compromised lives, bodies and molecular processes of the birds themselves.

The ready-made ecology of factory-farmed meat draws attention to a quite different economy of infectious disease. Disease is said to be endemic to such intensive environments, ever present as a risk by virtue of the agro-ecological set-up. Biosecurity measures, designed to prevent the passage of disease from the outside, in effect turn their back to (or even exacerbate) the incubation of disease within. Combined with the ecologies of production that seem to scale or 'bloat' everything from poultry populations, to chicken bodies as well the extended system of industrial regulation, there is clearly a need to investigate not simply the existence of this scaled-up agriculture as if it alone were the source of the problem, but the intra-actions or relations that generate the disease situation or the conditions for pathogenicity. Ecologies of production are critical, of course, but only, we would argue, in the sense that they are part of a relational situation where commercial and regulatory pressures are folded into a mutable world of hosts, microbes and immunological stresses with variable outcomes.

Relational Economy of Disease

Infectious disease, as previously argued, is neither a property nor an inevitable consequence of a given microbe (Casadevall and Pirofski, 2014; Lavau, 2013; Methot and Alizon, 2014), or for that matter the outcome of a particular industrial farm setting or factory. Instead, we have argued that it is a contingent outcome of particular *configurations*, in this case, of microbes, animal bodies, agricultural equipment, farm labourers, feed additives, and other materials; configurations that are generated in practices of animal and food production, as well as related regulatory, commercial and labour processes.

In labelling this as a relational *economy* of disease, following Barad (2007), we are concerned, as signalled in earlier chapters, with the significance of these sets of 'intra-actions' in producing disease. It is not just mixtures of things that are produced through these practices, but also the things themselves, for example microbes and animal bodies. Barad uses the concept of 'intra-action' to emphasise that such encounters are not between ready-made things, as suggested by the term 'interaction', but rather that these things are constituted in these encounters.

There is a sense in which the bundle of things that are jumbled together in poultry houses, for instance – birds, equipment, microbes, feed technologies, contract labour – are not reducible to so many discrete objects that merely interact. As argued in Chapter 3, microbes and hosts do not just bump into one another as separate, pre-determined entities to produce an effect called disease; they intra-act, that is, they work through one another to co-generate something that was not necessarily in evidence before, and is not simply reducible to a combination of pre-existing entities. Disease, when seen in this light, is the outcome of a *continuous interplay* between animals, microbes, people and materials that intra-act as they circulate, producing a shifting and a more or less pathogenic landscape. Disease, as such, is thus neither the result of some predetermined, discrete causal mechanism located within the pathogenic agent, nor indeed necessarily of a particular factory or farm setting.

As Barad (2007) suggests, such intra-actions are not bounded by the constraints of a particular place or time, but rather take their shape topologically, cutting across or rather dissolving any simple inside/outside diagrams as noted in the previous chapter. There are, of course, physical demarcations between the inside and outside of factory farms or processing plants, but the already entangled phenomena take their shape as much from the world of commercial and regulatory pressures *beyond* the confined sites, as they do from more immediate encounters within. Agricultural routines and materials of animal standardisation, together with growth acceleration, stock concentration, and confinement are embodiments of commercial pressures from further up the supply chain, immediate in their impact, yet distant in terms of their leverage. In particular, the imperative from retailers to produce just the right number of birds at the right weight and price point at the right time is, as we shall describe in detail below, also part of a relational economy of disease. Such commercial pressures are not 'external' to disease. They too may be constitutive of disease, lifted out and folded in such a way that they topologically re-embed themselves in a series of intra-actions that generate disease situations including, in different ways, those of *Campylobacter* and Avian Influenza.

Commercial pressures, such as the demand for just-in-time production, are thus not something experienced at one remove in modern factory farming. It is not simply that such conditions inevitably or naturally produce factory environments which harbour disease, but rather that to produce just-in-time birds has ramifications for much of what enters the relational mix: the growth regimes installed, the stocking densities, the medicinal regime, the timetable for clearing the sheds for slaughter, as well as the regulatory requirements for animal welfare and practices of contract agricultural labour that make up the configuration. These intra-actions affect pathogenicity in ways that are not always easy to read from the off. Importantly, it is difficult to reduce the situation to one or several inputs, so attempting to address the *Campylobacter* situation by focusing on the bacteria alone is insufficient, and likewise the focus on accidents or departures from the corporate script may be a diversion too. Disease situations are multi-factorial, and subject to the intra-actions and changing dynamicity that we detailed in previous chapters.

A *Campylobacter* Situation

Although *Campylobacter* grows well in chickens, it does not grow so well in culture and was not successfully isolated until the early 1970s. The difficulty in culturing this bacterium has delayed and complicated attempts to characterise its physiology and ecology, to develop reliable tests for its presence, and to identify infection pathways. As one major UK poultry grower admitted, the relationship between trialling a biosecurity intervention in a poultry shed and the flock subsequently testing green (negative) or red (positive) for *Campylobacter* seems somewhat random. *Campylobacter* also surprises in terms of its rapid emergence. A typical shed of 30,000 birds can become positive for *Campylobacter* virtually overnight. As one major UK poultry processor observed:

> Typically if you've got broilers, say at 40 days, it can be clear of *Campylobacter*, and then, within 24 hours, the whole flock can be [positive]. (Interview)

Campylobacter seems to defy attempts to map or control it, and there is a degree of frustration within the poultry industry about not being able to pinpoint any clear biosecurity breach or indeed identify what triggers an outbreak. *Campylobacter*, in the broiler shed, is perhaps less an outbreak and more a potential or ever present and so 'virtual' threat. It is something that always shadows the process of food production and is difficult to disentangle from the various elements that come together within industrial chicken sheds and factories. What is widely acknowledged within the poultry industry, however, is that there seems to be an association between *Campylobacter* and the process of 'thinning' the flock, which keeps the density of birds within a shed within animal welfare regulations, whilst maximising the profitability of the shed space (Allen et al., 2008; Food Standards Agency, 2010a).

The process of thinning brings a variety of different elements into play, so that cheap protein, pathogens, profit margins, just-in-time delivery, stocking densities, subcontract labour, and animal welfare standards all come together in what appears to be an uneasy co-existence. Poultry production on an industrial scale, as we have had cause to stress, leaves little room for manoeuvre in already tight production schedules when the profit margin per bird is so low. Thinning flocks is deemed essential to maintaining the industry as a viable business. As one manager of a processing plant explained:

> From an economic perspective, within this country if we were to stop the thinning that may be a problem. That may help as a solution [to *Campylobacter*], but from an economic perspective it's just not achievable. We wouldn't be able to have a poultry industry in this country if we weren't allowed to thin and grow birds, because of best use of buildings and the land and the economics surrounding that. At the moment, the profit per bird is very, very small... (Interview)

Chicken sheds are stocked at densities (kg/m^2) that have to comply with animal welfare regulations. As the birds put on weight, the poultry houses are not only at risk of breaching the legal limit, they are also in danger of growing beyond the right commercial price point. Thinning flocks, that is the removal of 20 to 25% of the stock when the birds are roughly 35 days old, keeps the bird density in check, while the remaining birds are left to grow on to the next desired price point. The effect is to raise the productivity of the sheds as a whole, in terms of poultry meat per square metre, by effectively producing two crops of birds each cycle, at two different price points. In the end it all comes down to the growth rate of the birds and when they hit the desired price point, with each retailer dictating its own product specifications down the line (e.g. one retailer specifies thinning at 1.9 kilos and a kill at 2.2 kilos).

However, whilst such practices do indeed make sense in commercial and animal welfare terms, there is research to suggest that the stress of the thin may leave the remaining birds vulnerable to opportunistic colonisation by *Campylobacter* (Humphrey, 2006), whether already present within the flock or having been introduced during the thin. Stretched by accelerated growth cycles, and at a life stage when their immunity is at its least effective, such avian bodies may tip from healthy to diseased, and indeed, the catch itself may it is argued prompt the uptake of harmless gut *Campylobacter* into muscle where it can, in the right conditions, be harmful to consumers (Humphrey et al., 2007):

> Another challenge I think is control at thinning. You've got stress of the birds at that time. I think that probably reduces their resistance and also you're then introducing catching teams, so the hygiene of the catching teams, equipment as well is coming in at that stage. So it's not so much the birds going out, but it's the birds that are left which are the ones that are at risk of infection. (Interview with major UK poultry processor.)

As an industry expert suggests, this process has effects that are not only about infection, but about the dynamics of bacteria and hosts:

> That is because of the stress of catching, being put in the crate, being starved, and one of the things that happens (…) their gut is flooded with noradrenaline (…) [the] noradrenaline captures iron and takes it to *Campylobacter* and increases its growth rate by about 10-fold. That, we think, explains the difference between transported and non-transported animals, and because the iron that the bug has now got has switched on virulence genes, this bug, if you give it another chicken, is much more invasive. (Interview with industry scientist.)

Pathogenicity is borne then of the situation. Equally, however, it is not the thinning process itself that is at issue here, but rather the different elements of the configuration that come into play in the act of thinning, one of which, as mentioned above, is the introduction into the mix of the catching teams. In a system where everything it seems has been scaled up, there is one factor of production that has been scaled down; labour. Most chicken farms run with a skeleton staff of perhaps one full-time manager and a part-time stock person. But thinning and 'harvesting' are labour intensive and in the UK system require just-in-time hiring of labour to catch and remove birds. Catching teams, and companies, are often subcontracted for this work and make money by operating over a wide region, which can include for example the South West, or even at times of high demand, the whole of the South coast, Northern England and so on. The teams, usually all male, are hired often on a piecework basis and can work long shifts on multiple sites in order to make sure the birds reach the processors on time. It falls to them to keep the weights, price points and densities in balance by moving from farm to farm to clear the sheds or to thin a flock. The catchers' work is choreographed by the processor through what is colloquially known as the factory's 'kill plan', or more euphemistically referred to as the 'daily collection schedule', so that trucks, forklifts and modules, catching teams, and fully grown birds come together at just the right time. There is little if any slack in the system, so these teams take the pressure and are hired at short notice in order to maintain the flow of chickens to the processor.

Precarity and Intra-Active Lives

Catching is physically demanding work performed in humid and 'stupidly hot' conditions, as one catcher described it, often in darkness to keep the birds relatively calm. Workers are given protective clothing that is cumbersome to move around in, and so is often discarded or left outside the shed. They work on a month-old substrate of chicken litter (often 30–40 cm thick), all the time breathing poultry dust, which invariably contains an unhealthy combination of fine particles and microorganisms. Few catchers wear respiratory masks as they quickly become ineffective in the humidity and are uncomfortable to work in – like the overalls they are often discarded in the rush to meet the contractual obligations and deliver the chickens just in time. Birds are 'collected' handfuls at a time, with a 'handful' dependent upon what exactly is specified in the contract or farming system. A common practice for 'standard' chickens is for catchers to slide the chicken's claws between their fingers, turn the birds upside-down, up to seven sometimes eight birds at a time (four on one hand, three on the other), and deposit them in the module trays left by the fork lift truck drivers, filling the drawers one by one. The chickens may be more or less docile, depending on the status of the flock and the conditions. But workers are used to chasing birds, to feathers flying, to being pecked and scratched.

As a rule, it takes a catcher a couple of minutes to fill a drawer with 24 birds. Each module usually has 12 drawers, making a total of 288 birds per module, and an average lorry carries around 22 modules, adding up to over 6,000 birds per lorry. It takes around 45 minutes to an hour for a team to complete a lorry's load, with teams over the course of a shift catching tens of thousands of birds, often across more than one site, depending on the requirements dictated by the processor. The catchers know these numbers well. They have to keep up with them, for these are the numbers that feed the high throughput at the plant, which, in turn, helps to keep the price of chicken low. The lorries are pre-ordered and need to return on time, and full with the requisite numbers of birds. The margins are so tight that lorries returning without a full load or indeed the need to send an extra truck to remove birds if the counting has not been done properly will jeopardise the profits.

The majority of poultry catchers in the UK are employed on a contract basis by the processor and they occupy a niche role in the large-scale industrialisation of poultry production. Some processors do employ their own in-house catchers, but the contract relationship suits the dynamics of a tightly coupled integrated system as it offers the processors a degree of flexibility over the timing of the catch, one that coincides with the not entirely predictable growth rate of the birds and their price points. Whilst the logistics of the catch are planned well in advance to match the growth cycle of the birds, the schedule is not finalised until the last minute, so if any delay occurs, such as a breakdown in production at the plant, the catching team absorbs the pressure. There is little slack in the system and time has to be made up to ensure the throughput of numbers. From the processor's point of view, the birds must arrive in time to keep up with the pace of the production line and workers. But the birds must not arrive ahead of time at the factory, as they may unnecessarily suffer or die inappropriately. The birds, having achieved a specific weight, must arrive at the plant just in time.

Crucially, a small change in the pace of collection or in the schedule itself can have a knock-on effect, not only on profit margins, but on the relationship between human and avian bodies that are already living and working in intense conditions. The precarious lives of the poultry catchers, often casualised and without sick pay, where illness (diarrhoea and farmer's lung are the most common) is an occupational hazard, intra-act with equally stressed avian lives. The fast growth of the birds, genetically designed to put on lots of flesh as rapidly as possible, but not matched by increased leg strength or heart and lung capacity, leaves them compromised in stressful situations that the arrival of a catching team and their imperative schedules foreshadow. Indeed, the different stresses experienced by compromised birds and casualised labour, their combined precarity, may itself tip a disease situation. Given the intra-actions of just-in-time poultry, casual labour and stressed avian lives, poultry workers and catchers in particular become a front line in bacterial and viral chatter (Chapter 3). Many workers develop enteric illnesses in their first few days at work, but have little or no incentive to stay off work in this casualised business. Workers also have been known to develop antibodies to avian influenza viruses which confirm that these precarious bodies, despite all the regulated exchanges, are in touch with one another and intra-acting (for a review of evidence and the need for poultry worker protection, see MacMahon et al., 2008). The tendency for catching teams to employ migrant labour, many of whom may have little or no direct access to health services or be registered with a General Practitioner, and may well be living in substandard accommodation, only exacerbates the potential problem (Lawrence, 2015; Liebler et al., 2009).

Whether or not disease is subsequently detected in the flock, however, is contingent upon the relational economy of the thinning; that is, the production practice that holds together such diverse elements as the logic of cheap chicken, commercial pressures on the factory, regulatory demands about the treatment of animals, contract labour practices, and the growth cycle of chickens. But it is a contingency that shows no clear pattern. The intra-actions of stressed chicken bodies, human bodies, microbes, and so forth may, or may not, produce a diseased flock. It is not that one change in isolation will automatically tip a flock over the threshold, but rather that when such diverse pressures are folded into the volatile situation, the potential may be there for such a critical point to be reached. Pathogenicity, as argued in Chapter 3, is a product of such intra-actions and as such not reducible to any one element. Even if the infection pathways and the behaviour of the bacteria remain unclear, the thinning process carries with it an association with *Campylobacter* and is indicative of a tipping point, a moment where an already compromised life is open to ever greater risk.

Powers of Life

In conventional accounts, the power exerted over the poultry supply chain, and indeed over life itself, by the big commercial companies and supermarkets is an instrumental exercise oriented towards one end: the commodification of poultry as factory farmed meat and the application of genetic and pharmaceutical technologies to produce birds that grow just-in-time to the right shape and price point. The attempt to control for future risks is a part of that process, but such control also points to one of the *paradoxes* of such straightforward instrumental power: that the exercise of power over life to just the right weight and size, when effective, may actually work against itself. Following Roberto Esposito (2008) and foreshadowed in Chapter 1, the very attempt to manage life, to make it safe, has the potential to open up unanticipated and unforeseen threats where a different kind of power, that of the powers *of* life, may turn against itself.

The tight specification of the poultry stock, the application of pharmaceuticals to manage behaviour and stimulate growth, the optimisation of a particular body shape and the rolling out of a strict feed regime, all have as their aim the economic standardisation of life, its correction as it were, so that risks to the supply chain or the dangers of infection no longer readily pose a threat to profitability. But, by commerce taking charge of life in this way, that is, by stripping it down to the bare minimum required for the production of cheap animal protein, life is also reduced to the edge of 'living'. For factory-farmed birds in tightly cramped, stressed conditions, their compromised immune systems, as we have seen, occasion shifts in the diagram of disease and of the disease situation. The threat comes from within as much as without, and the disease situation starts to shift from an emergency of incursion to one that is more ecological. As such, the very attempt to make the life of the birds safe from an outside world of disease may have quite the opposite effect, and flip over into a pathological threat.

On this view, too tight a control over life risks reversal into the production of harms that hitherto were unknown or not perceived as a threat. In taking charge of the farmed animal population and reducing it to 'mere' life (Honig, 2009), the powers of life, the potential for its pathogenic proliferation, may actually be turned against itself. This account, then, whilst recognising the role that corporate and regulatory powers play in an attempt to secure life, goes a step further to argue that such regulation and control runs the very risk of an economically standardised and reduced life undermining the very life that it seeks to protect. It amounts to an argument drawn from Esposito that foregrounds the powers *of* life folding over into the power *over* life and undermining its very possibility. It is the protection from threats which emanate from outside of the factory farm system that runs the risk of exposing animals and humans to ever greater harm. Securing life by enclosing it behind barrier systems, one of the aims of modern factory farming, in this sense may actually risk a reversal into ever greater insecurity.

The view that disease risk is embedded within the factory farming system is one that we broadly endorse, but the account of power that often underpins such a view, that of a system with a centre that is somehow rotten, offers a somewhat excessive confidence in the capabilities of instrumental power. The sense in which instrumental power, the power over life, exercised by the retail corporates reaches beyond itself when faced with the contingent economy of just-in-time pressures and immunitary stresses, reminds us of its limits. It also reminds us that the powers of life, the possibilities afforded by life's excesses, are part of an arrangement for which the logic of more control may not be the answer. Tighter, more extensive controls over the lives of farmed birds and a scaled-up biosecurity system designed to prevent incursion may prove self-defeating in that they also produce better conditions for disease incubation.

Esposito's (2006, 2008, 2011) line of argument crosses over from the legal and political to the biomedical and, of course, is wider ranging in its implications for securing lives in the face of a multitude of risks. Here, however, we have drawn upon it to show the paradox of instrumental power and its uses in producing cheap factory farmed meat. That paradox, we would argue, runs through much of modern factory farming and its practices, and takes different forms in different situations. In the case of the poultry industry, when understood though the lens of a relational economy of disease, it takes something like the just-in-time pressures behind the process of thinning poultry to reveal the limits of commercial domination over life and its exploitation.

Conclusions

The aim of this chapter has been to draw attention to the diverse array of just-in-time pressures, both commercial and regulatory, that are folded into factory farming, arguing that these are constitutive of, rather than external to, disease risk. The pressures themselves, whilst the result of attempts to produce cheap, disease-free meat, contribute towards the very configurations that put at risk the lives of those that the tightly integrated systems were designed to protect. In that context, biosecurity measures enacted to keep disease out may not only wrongly focus on what anthropologist Sarah Dry (2010: 41) calls a 'fast twitch' or 'acute' approach to a more 'slow twitch' problem or situation, it may also inadvertently contribute towards the incubation of disease within and run the risk of ever greater pathological insecurity. When the well-honed inside/outside diagrams of biosecurity work as planned, in terms of disease, as we have seen in the case of *Campylobacter*, they may well work against themselves.

There is, however, nothing settled about a given agro-ecological system that suggests that disease is an inevitable outcome of the commercial and industrial systems deployed. As we see it, ecologies of production are part of a relational economy of disease, in which disease is a *contingent* outcome of the intra-actions among microbes, animals and profits, as well as commercial, regulatory, farming and food production practices. As important as it is to acknowledge the incubation of something like *Campylobacter* within a factory farm setting, it is equally important to recognise that disease is not simply a matter of bacterial or viral presence; it is a relational achievement, one that is generated through the entangled interplay of environments, hosts and microbes, not all the elements of which are present in any particular setting. It is, as we have argued, a disease situation; one of topological intensities and pressure points that have the potential to 'tip' a stripped down life over on itself.

What is also drawn into question by such a relational economy is our understanding of how power is expressed in the practices of the food and farming industry. The exercise of commercial and corporate power, not just over the food supply chain, but over the very liveliness of life itself, has had, we believe, the paradoxical effect of reducing life to a biological threshold in an attempt to control for future risks. When brought to a certain threshold, attempts to secure life, to protect it, following Esposito's line of thought, may actually end up negating it, with life transgressing all attempts to control it. It is in that sense that the powers of life have to be grasped in relation to the more familiar instrumental powers over life, which tend to dominate accounts of the disease risks evident in today's corporate agri-business and gloss over the non-human elements at its core.

References

Allen, V., Gittins, J. & Edge, S. 2008. *Evaluation of Best Practice Recommendations to Reduce Campylobacter Incidence Associated with Thinning of Broiler Flocks*. Final report to the Food Standards Agency. ADAS and University of Bristol, December.

Barad, K. 2007. *Meeting the Universe Halfway: Quantum Physics and the Entanglement of Matter and Meaning*. Durham, NC: Duke University Press.

Boyd, W. & Watts, M. 1997. Agro-industrial just-in-time: The chicken industry and postwar American capitalism. *In*: Goodman, D. & Watts, M. (eds), *Globalising Food: Agrarian Questions and Global Restructuring*. London: Routledge.

British Retail Consortium & British Poultry Council 2010. *Tackling* Campylobacter: *A Commitment Across the Supply Chain*. http://www.brc.org.uk/brc_m_policy_content. asp?iCat = 46&iSubCat = 482&spolicy = Food&sSubPolicy = Microbiology Accessed 17 August 2015.

Casadevall, A. & Pirofski, L.-A. 2014. Ditch the term pathogen. *Nature*, 516, 165–166.

Crane, R., Davenport, R. & Vaughan, R. 2012. Farm Business Survey 2010/2011: Poultry Production in England. *Rural Business Research, University of Reading*. Reading UK: University of Reading.

Davis, M. 2005. *The Monster at our Door: The Global Threat of Avian Flu*. New York: The New Press.

Defra 2015. *Poultry and Poultry Meat Statistics*, June 2015, London http://www.defra.gov. uk/statistics/foodfarm/food/poultry/

Defra, Northern Ireland Department of Agriculture and Rural Development, The Scottish Government Rural and Environment Research and Analysis Directorate and Welsh Assembly Government, the Department for Rural Affairs and Heritage 2010. *Agriculture in the United Kingdom 2010*. http://webarchive.nationalarchives.gov.uk/20130123162956/http:/www.defra.gov.uk/statistics/files/defra-stats-foodfarm-crosscutting-auk-auk2010-110525.pdf Accessed 15 August 2015.

Dry, S. 2010. New rules for health? Epidemics and international health regulations. *In*: Dry, S. & Leach, M. (eds), *Epidemics: Science, Governance and Social Justice*. London: Earthscan.

Esposito, R. 2006. Interview: Roberto Esposito. *Diacritics*, 36, 49–56.

Esposito, R. 2008. *Bíos: Biopolitics and Philosophy*. Minneapolis: University of Minnesota Press.

Esposito, R. 2011. *Immunitas: The Protection and Negation of Life*. Cambridge: Polity Press.

European Food Safety Authority & European Centre For Disease Prevention 2016. The European Union summary report on antimicrobial resistance in zoonotic and indicator bacteria from humans, animals and food in 2014. *EFSA Journal*, 14.

Food Standards Agency 2004. *Cleaner Farms, Better Flocks: Keeping Out Disease and Harmful Bacteria*. http://www.food.gov.uk/multimedia/pdfs/publication/betterflocksleaflet.pdf Accessed 17 August 2015.

Food Standards Agency 2006. *Biosecurity for Housed Broilers*. http://www.food.gov.uk/multimedia/pdfs/publication/biosecurityforbroilers1007.pdf Accessed 17 August 2015.

Food Standards Agency 2010a. *The Joint Government and Industry Target to Reduce Campylobacter in UK produced Chickens by 2015*. http://www.food.gov.uk/multimedia/pdfs/campytarget.pdf Accessed 15 August 2015.

Food Standards Agency 2013. *A Refreshed Strategy to Reduce Campylobacteriosis from Poultry*, September 2013 http://www.food.gov.uk/multimedia/pdfs/board/board-papers-2013/fsa-130904.pdf Accessed 17 August 2015.

Godley, A. & Williams, B. 2008. The chicken, the factory farm and the supermarket: The emergence of the modern poultry industry in Britain. *In*: Belasco, W. & Horowitz, R. (eds), *Food Chains: Provisioning, from Farmyard to Shopping Cart*. Philadelphia: University of Pennsylvania Press.

Godley, A. & Williams, B. 2009. Democratizing luxury and the contentious 'invention of the technological chicken' in Britain. *Business History Review*, 83, 267–290.

Graham, J.P., Leibler, J., Price, L., Otte, J., Pfeiffer, D., Tiensin, T. & Silbergeld, E. 2008. The animal-human interface and infectious disease in industrial food animal production: Rethinking biosecurity and biocontainment. *Public Health Reports*, 123, 282–299.

Hinchliffe, S., Allen, J., Lavau, S., Bingham, N. & Carter, S. 2013. Biosecurity and the topologies of infected life: From borderlines to borderlands. *Transactions of the Institute of British Geographers*, 38, 531–543.

Honig, B. 2009. *Emergency Politics: Paradox, Law, Democracy*. Princeton: NJ: Princeton University Press.

Humphrey, T.J. 2006. Are happy chickens safer chickens? Poultry welfare and disease susceptibility. *British Poultry Science*, 47.

Humphrey, T.J., O'Brien, S. & Madsen, M. 2007. *Campylobacter* as zoonotic pathogens: A food production perspective. *International Journal of Food Microbiology*, 117, 237–257.

Irvine, R.M. 2015. A conceptual study of value chain analysis as a tool for assessing a veterinary surveillance system for poultry in Great Britain. *Agricultural Systems*, 135, 143–158.

Lavau, S. 2013. Viruses. *In*: Adey, P., Bissell, K., Hannam, P., Merriman, P. & Sheller, M. (eds), *Routledge Handbook of Mobilities*. London: Routledge.

Lawrence, F. 2008. *Eat Your Heart Out: Why the Food Business is Bad for the Planet and your Health*. London: Penguin.

Lawrence, F. 2015. Fear, hunger and dirt: Lithuanian migrants on life as chicken catchers. *The Guardian*.

Liebler, J.H., Otte, J., Roland-Holst, D., Pfieffer, D.U., Magalhaes, R.S., Rushton, J., Graham, J.P. & Sibergeld, E.K. 2009. Industrial food animal production and global health risks: Exploring the ecosystems and economics of avian influenza. *EcoHealth*, 6, 58–70.

Macmahon, K.L., Delaney, L.J., Kullman, G., Gibbins, J.D., Decker, J., Kiefer, M.J. 2008. Protecting poultry workers from exposure to avian influenza viruses. *Public Health Reports*, 123, 316–322.

McCloskey, B., Osman, D., Zumla, A. & Heymann, D.L. 2014. Emerging infectious diseases and pandemic potential: Status quo and reducing risk of global spread. *The Lancet: Infectious Diseases*, 14, 1001–1010.

Methot, P.-O. & Alizon, S. 2014. What is a pathogen? Towards a process view of host-parasite interactions. *Virulence*, 5, 775–785.

Morgan, K., Marsden, T. & Murdoch, J. 2008. *Worlds of Food: Place. Power and Provenance in the Food Chain*. Oxford, Oxford University Press.

Perrow, C. 1999. *Normal Accidents: Living with High Risk Technologies*. Princeton and Chichester: Princeton University Press.

Red Tractor Farm Assurance, 2010. Red Tractor engages fully in the campaign against *Camplylobacter*. *Red Tractor Assured Chicken Production Newsletter*.

Wallace, R.G. 2009. Breeding influenza: The political virology of offshore farming. *Antipode*, 41, 916–951.

Wallace, R., Wallace, D. & Wallace, R.G. 2009. *Farming Human Pathogens: Ecological Reslience and Evoloutionary Process*. Dordecht: Springer.

Wrigley, N. 2002. Transforming the corporate landscape of US food retailing: Market power, financial re-engineering and regulation. *Tijdschrift voor Economische en Sociale Geografie*, 93, 62–82.

Yakovleva, N. & Flynn, A. 2004. *Innovation and the Food Supply Chain: A Case Study of Chicken*. The Centre for Business Relationships, Accountability, Sustainability and Society. Cardiff, Wales: Cardiff University.

Chapter Five
The De-Pasteurisation of England: Pigs, Immunity and the Politics of Attention

In Latour's telling at least, the *Pasteurisation of France*, and Pasteur's microbes, enabled a shift in powers (Chapter 2) as 'the scientist in his [*sic*] lab gets the edge over the local, devoted, experienced veterinarian' (Latour, 1983: 147). Germ theory, we have argued, re-inscribed a sense of inside and outside, and simultaneously not only shifted the location of expertise but also set in train a process by which the microbe-filled world outside the laboratory started to become more laboratory-like. Farms became, in this telling, part of the network of science. They and the food industry in general became the laboratory spaces through which the world, in Archimedean fashion, was raised. The pasteurisation of France was, even without a central Napoleonic hero, a progressive narrative. The germs of germ theory were matters to be extradited from the spaces of production. The diagram of disease was once again inscribed with a powerful division between insides and outsides.

In this chapter, we argue that such divisions are rarely so clear-cut. Insides and outsides, but also experts and non-experts, depend upon one another and are in that sense joined together. Indeed, our argument is that we require a rather different diagram or spatial imagination to understand pathological lives. We do so by looking at contemporary pig farming, aiming not only to trouble any hard and fast division between healthy and unhealthy living, but also to highlight the (still) vital role of the 'local, experienced and devoted' practitioners, and the role of microorganisms, in making health possible through a patchwork of practices

Pathological Lives: Disease, Space and Biopolitics, First Edition. Steve Hinchliffe, Nick Bingham,
John Allen and Simon Carter.
© 2017 John Wiley & Sons, Ltd. Published 2017 by John Wiley & Sons, Ltd.

(Law, 2004b). Without wanting in any way to undermine the vital role of pasteurisation in making safe lives, we nevertheless seek to balance this germ-free view by highlighting the importance of microorganisms in the making of *healthy* as well as diseased bodies. With a nod to the title of Latour's famous book, we talk of the potential depasteurisation of England. Following on from Chapter 4, we open up a discussion of the role that immunity can play in the production of alternative approaches to the emergency of emerging disease. Our question might be put this way: how do the very actors and actants (the farmers, vets, pigs and microbes), who seemed to be displaced by the laboratory scientist, contribute to healthy lives? And how might we best value those contributions?

We start with an introduction to pig farming, with the 'birth of the sty', looking at recent changes to pig farming, current market relations and the ensuing disease situation. We then discuss, in the section 'Pigs in Practice – Fieldwork and Translation', our empirical work with pig farmers and vets, noting how pig movements are changed, at least in theory, to more regulated circulations. On the farm, and in practice, things may not be so clear cut. We detail how claims to immaculate, germ-free agriculture are dependent on a series of practical translations that involve bugs, bodies and specificities, and re-introduce movement, contact and communication (see Chapter 3 for the distinctions between movement and exchange, etc.). The result is that farmers and local vets, and pigs and microbes, become as important to health as the scientists in their labs.

We end the chapter with the section 'Immunity, Attention and More-than-Human Responses', with an extended discussion of how immunity and the practices we have witnessed have consequences for the disease situation. We seek to mark out a distinction between the responsibilisation of practitioners (serving a biopolitical end) and the need for more responsive human–non-human animal relations. While the latter can resemble the 'ethicised states of being' and continual vigilance that seem to pervade prevailing modes of security and emergency politics (Adams et al., 2009), we nevertheless attempt to rescue a responsive mode of living pathological lives from current versions of biopolitical subjectivities. Our overall aim is to mark out an area, for debate at least, on what it might take to depasteurise England.

Birth of the Sty

In order to understand the pig situation, or pathogenicity in pig farming, and the openings that there may be for doing things otherwise, it is useful to provide a brief overview of current farming practices, market relations and pig diseases. These are clearly interrelated matters, part of the broader situation, but we will deal with each in turn.

Into the Indoors

As Dawn Coppin (2003) writes, pigs and people have co-evolved since the earliest domestication of wild boar around 10,000 years before present. But only in the last few decades have we seen the rise of the mega-farm, with herd averages in the US of over 5,000 pigs per farm. Coppin's characterisation of these single species farms with their accelerated throughput of pigs is worth quoting in full:

> A typical hog operation today, involved in all the stages of swine raising from breeding to finishing hogs to market weight, is a quite different place from the farm fifty years ago. Breeding is mostly done through artificial insemination, although estrus may be detected either by live boar, a human, or a machine. Once pregnant. the sows are kept in an individual crate for about three and a half months until a few days before they are due to farrow. At this time they are moved to other individual crates where sows give birth and where they and their litters stay for two more weeks until the piglets are weaned. The piglets are then moved to a nursery building for a month, then to a growing building, and finally to a finishing building where they stay until they are five to six months old at which time they are loaded onto a semi-truck to go to slaughter (Coppin, 2003: 600).

As Coppin details here and elsewhere, the evolution of this system of production has involved the delegation of powers to material arrangements and infrastructures. For example, in the twentieth century in North America, pig production and pigs moved indoors, as they did in parts of northern Europe (Geels, 2009). There was a gradual confinement of pigs, and particularly of sows (pregnant female pigs), that involved decades of on-farm experimentation and a new discipline of animal science. Indoor pigs were potentially more efficient producers of pork meat, but moving pigs indoors was not economically straightforward as they needed to be fed, kept clean and exercised, jobs that potentially involved extra human labour.

Moving pigs indoors in order to improve productivity and reduce mortality necessitated further invention in order to solve new problems. So the indoor farrowing pen, itself introduced in order to reduce mortality in piglets born outdoors, set off what Coppin calls a round of 'human–porcine–material interactions' (Coppin, 2008: 52). While young piglets were now protected from the vagaries of weather (especially Mid-Western extremes), they nevertheless started to suffer from new conditions related to being confined indoors. Infectious diseases became more common and were soon linked to hygiene standards within the pens. Slatted floors were subsequently introduced to facilitate drainage of excreta, and straw bedding was often removed from the pens in order to reduce microbe-rich materials. The resulting strawless pen with slatted flooring and concrete under-floor also reduced labour inputs, in the form of mucking out or scheduling pigs to do their business outdoors. But the resulting pens now needed to be heated, as stationary nest-free bodies were poor at temperature regulation. And the tendency of mothers to roll onto their young in relatively cramped and hard surfaced enclosures led to further confinement of sow bodies as pens became crates, surrounded by a space for piglets to move away from their mothers. Likewise, new feed and drug regimes were necessarily adjusted to these newly docile bodies. In one example that has resonances for later in this chapter, farmers used to place a sod of earth into the pens as a way of reducing illnesses in piglets. It turned out that the success of this folk cure was in part at least based on the trace amounts of iron in the soil. Piglets with access to soil were less likely to develop anaemia and secondary illnesses. However, the soil was soon suspected of bringing diseases and parasites into the pen, and in time it was replaced with expensive iron supplements and injections.

With these stories of continual and incremental change, Coppin's material history of the North American sty aims to draw attention to the short and arguably impoverished lives of pigs. It also aims to unsettle the sense that there is anything determined about the evolution and current arrangements of human–pig relations. Things may, as science and technology studies often tells us, be otherwise.

In the UK, the mega-farm is a relatively recent arrival, spurred on most recently by discourses of food security and sustainable intensification (Parliamentary Office of Science and Technology, 2012). A recently planned though ultimately rejected development in Foston, Derbyshire, for example, would house 2,500 sows and produce 1,000 pigs per week for sale, holding around 25,000 pigs on site at any one time (Midland Pig Producers Ltd, 2012)[1]. While farms of this scale are a rarity in the UK, the bulk of British production is indoors, and nearly all piglets are 'finished' (that is fattened up for slaughter) indoors (BPEX, 2013; DEFRA, 2014)[2]. According to some, and despite a small outdoor rearing sector and a somewhat niche organic sector, there is a growing tendency towards larger farms, even super farms, with production based entirely indoors (Parliamentary Office of Science and Technology, 2012). That said, divisions between farming systems are not quite so clear. The historian Abigail Woods (2012), for example, details the debates in the UK that occurred between indoor and outdoor pig producers in the early to middle twentieth century. She notes how outdoor and indoor production were not so much marked by a progressive narrative and a gradual displacement of more extensive systems by the more intensive. Nor were outdoor farms romantic reactions to creeping modernisation. Rather both forms co-existed in the UK, each responding to different market opportunities and to changing understandings of animal science. And both borrowed from each other and developed systems that combined practices and technologies. The same is to some extent true today, as outdoor sow herds have flourished in response to high fuel and feed prices, new animal welfare regulations (the 1999 UK banning of sow stalls, for example) and relatively low pork prices. At the same time, the vast majority of piglets born on these farms are moved indoors for 'finishing'. The point is that while there is a tendency for the proportion of indoor-based farming to increase, the sector may still be innovating in ways that combine practices and forms of farming.

Market Relations

Absent from these statistics and studies of the national characteristics of pig holding are the market relations within which pig keeping is practised. In a sector facing stiff competition from North European producers, and with rising costs of key inputs including feed, and the need to keep up with animal welfare regulations (some of which are more exacting than in other European states), farming pigs in the UK is often seen as a somewhat precarious living. Another dimension to the UK pig scene in particular is the '"opportunist dealing" right along the food chain which undermines efficiency and creates a dysfunctional, adversarial food supply chain' (Bowman et al., 2012: 6). In order to understand the pig disease situation, and pathogenicity in pig farming, it is important to explain how this opportunist dealing came about through the market relations that exist between retailers, processors and farms.

Similar to poultry (Chapter 4), the large supermarkets are key players in the sector. In the last decade or so, supermarket chains have responded in different ways to rising price competition, saturation in the domestic retail market and the need to deliver shareholder value[3]. But a more general feature of their response has been to adopt opportunist, and flexible 'supply agreements' instead of fixed-term contracts. The result, Bowman et al. (2012) argue, has been for retailers to play the three main suppliers of processed pig products off against one another in an oversupplied market. The processors (Vion, Tulip and Cranswick, which account for over 70% of UK market share) certainly enjoy good market relations with the retailers (often to supply own-label products). However, these relations are not based on long-term contracts. So while the relationships are consistent enough for both parties to establish reasonable market stability, market relations or the exact content of that relation is in principle precarious (supermarkets can walk away at any point). They guarantee neither price nor volume and can be renegotiated on a monthly basis. These renegotiations tend to involve a continual retailer-led press for lower prices, for raising promotional offers (with the costs passed down the chain to the processor and to the farmer) and a routine holding back of a portion of the agreement 'for tendering by other suppliers and main order switching' if processors fail to deliver on the 'latest demands' (Bowman et al., 2012: 28). This 'trader mentality' has effectively infected the entire sector, with the large multi-national processors offsetting any temporary losses in order to retain 'turn over, exclusivity and market share' (Bowman et al., 2012: 30).

Perhaps the most pernicious effect of this mix of powerful retailers and internationally owned 'super middlemen' processors able to take short-term financial hits, is what happens on the farm. As processors feel the pricing pressures of the supermarkets, farmers struggle to sell their product at a profit. Moreover, given the instability of the monthly renegotiations, there is a degree of volatility of price in this non-contract arrangement market. The result is that farmers also adopt a trader mentality, and 'commonly decide when to sell their animals based on the spot market price' (Bowman et al., 2012: 31). For example, a mid-sized indoor pig farm in Somerset that we visited sold their pigs through a marketing intermediary, which 'controlled thousands of pigs each week going into the slaughter houses' (Farmer interview). Despite this control and market power, the result can be last minute changes to slaughter numbers:

> So like this week I booked in 110 but we only sent 90 because they're cutting everyone back because of lack of demand etc, etc. (Somerset Pig Farmer, Standards)

This approach to sales, Bowman et al. (2012) suggest, results in farmers aiming to maintain a competitive edge over their peers and militates against collaborative or cooperative action. The precarity and lack of longer-term stability reduces investment capacity and the lack of cooperative or horizontal links between farmers diminishes collectively sanctioned improvement in production processes. The absence of incentives for investment and indeed the poor international competitiveness of the UK pig industry, a House of Commons Select Committee concluded, stem from this supply chain instability and lack of longer-term contracts (House of Commons Food and Rural Affairs Committee, 2009: 9).

Pig Diseases

Against this backdrop of uneven and far from straightforward change, and of market instability, there has been and remains a constant threat of disease within the pig sector. Pig production is troubled by a number of diseases that effect pigs and that can cross into people. Pigs are generally perceived to be impure by many of the world's religions and trading bodies, with pork meat traditionally being associated with parasitic infections like *trichinosis*, and subject to cultural norms around curing and cooking, and thereby a major target of EU food safety standards (Dunn, 2005). Pigs may be particularly susceptible to respiratory diseases and pneumonia as a result of a relatively small lung to body size ratio (itself a product of centuries of selective breeding). They are also particularly efficient at amplifying and shedding viruses like classical swine fever, foot and mouth and influenza. Indeed, the socio-economic impacts of classical swine fever in 2000 and foot and mouth disease in 2001 and again in 2007, which resulted in movement and export restrictions, have arguably contributed to the lack of investment in the sector (House of Commons Food and Rural Affairs Committee, 2009). Pigs are also regarded as a key species at the human–non-human animal zoonotic interface. For influenzas, they are thought to be the most likely 'mixing vessel' within which co-infections of various strains of influenza can re-assort and generate viable lineages that could infect both pigs and people (though see Short et al., 2015 for an update to this mixing vessel assumption). Recent phylogenetic characterisation provides strong evidence that the pandemic H1N1 'Mexican' swine flu outbreak of 2009, for example, involved a series of re-assortment events involving avian, pig and human origin influenzas, with pigs as a key host in the triple re-assortment pandemic swine virus (Smith et al., 2009). Finally, pigs are also key intermediaries and amplifiers for other zoonotic diseases including Japanese Encephalitis, and their guts commonly harbour *Salmonella*, *Campylobacter* and *E. Coli* bacteria.

Given this pathological life of the pig, the birth of the sty has involved an assemblage of disease abatement strategies at the same time as it has changed pathogenicities through increasing sizes and densities of herds. We have already mentioned the slatted floors and the removal of nesting and other palliative materials from some farms. Most farms will also operate 'all in all out' systems (similar to poultry) where a generation of piglets are kept together from birth to finishing in order to minimise mixing with other pigs. There is a routine use of antibiotics in order to control bacterial infections; sales of antibiotics destined for food animals in the UK are roughly similar in tonnes per active ingredient as for human use – and pigs and poultry account for 80–90% of that use (Veterinary Medicines Directorate, 2013) – in the US the figures are even higher owing to the non-therapeutic (preventive and growth promoting), use of antibiotics. In addition, herds are often spoken about as closed, meaning that they are, in theory at least, sustained through in-house breeding, farrowing, weaning and finishing. Pigs in this version of farming do not circulate between farms – they circulate on the farms through the various stages of life until being sold for slaughter. Finally, biosecurity is taken seriously, it is often said, in the pig sector. Partly because of the susceptibility of pigs to disease, most pig farms insist that visitors are pig-free, meaning they have not visited a pig operation or possibly eaten pork products in the previous few days – even vets and inspectors must observe this stand-down time as a precaution against transferring infectious materials. Any pigs that do circulate through the farming system are advisedly quarantined *if* farm infrastructure allows.

To sum up, the birth of the sty has involved a continuing choreography (Cussins, 1996) or dance of agency (Pickering, 2008) involving pigs, people, markets and microbes among many others. There is a story emerging here of how market relations, fluctuating costs and pigs as trans-species bodies contribute to a more or less pathogenic situation. To be clear, there is a tendency, as we have spelled out earlier, to argue for the continuing hygienic enclosure of pigs and other livestock in order to realise surplus value and to reduce pathogen incursion. But we have also started to unsettle this progressive narrative, hinting at the continuing importance, contra the pasteurisation story, of a range of actors and actants in producing health. While the laboratory scientist may have become a vital component of healthy life, we are not yet sure that they 'have the edge over' farmers, local vets and, indeed, pigs. In order to explore this further we turn now to our fieldwork and to pigs in practice.

Pigs in Practice – Fieldwork and Translations

Despite assurances regarding the levels of biosecurity that pertain in pig farming, in practice things are rarely so neat and tidy. In order to demonstrate how, we draw on fieldwork on farms and with veterinarians and others involved in the pig industry. The fieldwork took place in South West England (mainly in Devon, Dorset, Somerset and Wiltshire). It should be noted that the South West has become an area with a relatively high number of pig operations and pig numbers (DEFRA, 2006)[4]. That said, pig density is somewhat less than that found in traditional pig areas, suggesting many holders operate smaller herds and that larger herd sizes are more likely (but of course not necessarily) stocked at lower densities. A strategic sample of 20 holdings was selected for the study and a range of farms, organised by size, and type of production, were visited and farmers interviewed. Interviews were semi-structured and open-ended in order that we could make sure they covered the ground that we anticipated would be important but also allowed participants to tell us about and show us issues or arrangements that they considered of most importance. Where possible, interviews were followed by walking tours of the farm and a chance to meet the pigs and ask further questions about the operation and biosecurity arrangements. Field notes were kept of these site visits. We also interviewed veterinarians involved in the pig sector to discuss endemic and exotic diseases and procedures of disease control. Likewise, we talked to two breeding companies and to industry representatives (assurance schemes, and pig producer organisations) in order to understand the practices involved in the breeding process and to gain a broader view of the industry as a whole. All interviews and field notes were transcribed and analysed with the assistance of *NVivo* software. Analysis of all the ethnographic data involved identifying codes that summarised text fragments, with an iterative reading and re-reading of transcripts and a continuous process of review and re-evaluation of codes. Constant comparison methods were used to systematically refine these codes. This involved a process of simultaneously coding and analysing data in order to develop concepts. Continual comparison of specific incidents helped to refine concepts, identify their properties, explore their relationships to one another, and either integrate them into an explanatory model (Taylor and Bogdan, 1984: 126) or in many cases maintain their non-coherence in order to preserve some of the multiplicity and messiness of social and material practices (Law, 2004a). The point of this kind of research is of course not to claim that we have managed to access a numerically representative sample of UK pig premises. Rather, our in-depth engagements with practitioners on a range of farm types as well as interviews with key industry mediators (vets, breeders, assurance agencies) and analysis of those encounters provide confidence that we have generated significant insights that apply across the sector. We would add that, for this study, the range of South West holdings provided an interesting and useful sample through which we could explore, in depth, some of the practices involved in making healthy pigs. We now turn to these practices, looking at movements, circulations and translations in turn.

Pig Movements

For a small handful of pig diseases, farmers were somewhat fatalistic about their chances of avoiding a breakdown. Biosecurity has its limits it seems. Foot and mouth disease, which was widespread in the UK as recently as 2001/2, occasioned this kind of sentiment:

> Foot and mouth is a classic, because there's so many pigs/animals on one unit, in such a small area, [you] get a massive like mushroom thing, a nuclear cloud of a pig farm of foot and mouth disease and that just tumbles down to any neighbouring farms and gets spread that way. (Farm manager, Devon (freedom food pigs)

In these more extreme cases, the rituals of biosecurity, which include spraying truck tyres with viricide, seem symbolic rather than functional. Nevertheless, for most infectious diseases, movement is a key concern. Domestic pigs are most at threat in terms of infectious diseases from other pigs, from pig products or from other organisms, including people, who have had recent contact with pigs. Movement in this sense may be risky. But it is also something that may be required to make ends meet. Some of this relates to the market pressures we have already mentioned. One farmer, for example, talked about the time that pneumonia took hold on his farm (we will say more about this disease in a moment), when mortality rates suddenly increased and production rates declined:

> A bad year would be less than 2% [mortality rate], a good year would be less than 1% and that's weaning to slaughter, and some of those deaths would be fighting or something like that, you know... But towards the end of us producing the commercials we had a lorry come down that had coughing pigs on, and my pigs then got pneumonia and I could not get rid of it. (Farmer, Devon, Standards)

Under pressure to meet commercial targets and turn a profit, and with an eye on the fluctuating prices for pork, some farmers occasionally need to buy in extra stock for finishing. Even if this is done carefully, using on-farm infrastructure to keep populations separate and so quarantining incoming pigs, the result can be an infectious disease that quickly affects the whole farm. The ease of indirect fomite transfer (on boots, feed, tyres and so on), air- and dust-borne microorganisms, and other vectors (flies, rodents) all make isolation a continuous battle. And even isolated pigs remain eminently infectable.

In addition to the tendency for some farmers to mix stock in order to ride the volatility of UK market conditions, others take advantage of other market openings. For example, one farmer bought older breeding stock at livestock markets or directly from farms, in order to turn a quick profit by holding mixed stock prior to shipping for the German sausage market. Last but by no means least, stock mixing may also occur for the 40% of UK pigs that are currently outdoor weaned and/or reared and are often moved indoors for finishing. This may involve on-farm or between farm movements.

Given the risks associated with moving pigs, many commercial farmers and certainly most large indoor pig operations prefer to operate herd systems that are less open:

> …so we don't buy in, whereas [names another farmer] who we were selling the store pigs to, […] they'll go and they'll have store pigs coming in from three or four different farms, so each farm's going to be slightly different disease wise, then mix them all together, you could have a big breakdown. So we haven't got any other stock coming on the farm other than the gilts, and we know where they're coming from and what their status is… (Pig farmer, Somerset, standards)

As this interview extract suggests, there are two kinds of movement here – one that seems less regulated and involves mixing pigs from different farms, all of which will be 'different disease wise', and then buying in stock from a single farm in which disease status is known. The latter commonly involves purchasing pigs from breeding herds, the aim being to replace older sows and so replenish the breeding stock. For reasons that we will detail, and in the terms that we used in earlier chapters, this movement might be characterised as a more regulated circulation of pigs and pig material. Here, pig contacts turn into more tightly managed exchanges, and communication is, at least in theory, reduced to commercial transaction (after Mitropoulos, 2012). We will now detail how pig movements are turned, at least in theory, into this more regulated process of circulation, exchange and commerce.

Circulations

Even in relatively closed systems (where 'closed' refers to a system whereby replacement animals are produced on site) in order to prevent in-breeding and degeneration of the farm stock, new breeding stock or 'pig genetics' must be brought onto the farm in some form or other. Breeding boars, or gilts (young female pigs), may well be regularly brought onto a farm in order to replace breeding stock, or the sows whose productivity is measured in the size and viability of each litter, and who tend to suffer a fall off that on average occurs after around 6 to 7 litters when they are 3 to 4 years old. Replacement may be frequently arranged in batches. One farmer who ran a herd of 420 sows would take 12 replacement gilts a month in order to maintain productivity. Other farmers considered even this more regulated movement to be risky:

> The policy of this farm is to breed their own replacements, that is a big disease, biosecurity issue, if you're buying in gilts from a breeding company, you're constantly opening the door and therefore if the donor farm breaks down with a disease and it doesn't show itself and you've got the pigs in, they go 'Oh Christ, we've broken down', it's too late. (Farm manager, Devon, Organics)

So replacement pigs come either ready-born, to order, from breeding companies, or they are generated on site. The latter can involve on-site servicing by boars, or artificial insemination, with the latter growing in the UK in recent years, but still relatively low compared to the Netherlands and Denmark (Marquer et al., 2014). In all of this, breeding companies are key, as their boars supply the semen that is necessary to maintain the viability, genetically, of the herd. These breeding companies increasingly refer to themselves as genetics companies, in the business of selling genes and hereditary dynamics (Holloway et al., 2009) through replacement pigs, semen or even foetuses for implantation.

A closed herd in this sense is a relative rather than absolute term. So, a closed system today usually involves selecting breeding stock from the in-house generation of piglets, using semen or boars from a breeding company to produce a new generation of breeding and slaughter stock, and/or ordering a new generation of gilts to replenish the breeding generation. The breeding company will often be involved in advising farmers and their vets on genetic lines and breeding programmes using a BLUP or 'best linear unbiased predictor' model. The latter makes use of a vast dataset generated on commercial and breeding farms in order to manage inheritance, guard against in-breeding, and maximise productivity and other desirable traits (including disease resistance, hardiness for outdoor breeds and so on).

In the terms of the birth of the sty, this regulated circulation of replacement breeding stock is a relatively recent development. Pig breeding became established as a professionalised service in the 1960s, making use of an accumulation of decades of farm data and the operation of breeding pyramids (Woods, 2012). In order to 'guarantee' the health status of these circulating pigs, the 'nucleus' or 'elite' breeding herds are located away from other pig populations. They are promoted as remote 'pig-free' locations, with no or few registered pig operations in the vicinity, to minimise risk of airborne or vector-borne diseases. The aim is to produce bio-secure and disease-free pigs for circulation along highly regulated pathways to commercial producers. Remoteness sells in this line of work, and companies make a significant play of their location and origins in relatively bio-secure environs like Ireland and Canada:

> The fact that Ireland is an Island Nation, surrounded by sea, provides a natural barrier to the movement of animals which in turn benefits the biosecurity of the National pig herd. (Breeding company website http://www.hermitagegenetics.ie/ health/index.php, accessed 20 February 2015)

In turn, progeny from the nucleus herds are crossbred on separate sites called multiplication units; for one of the UK breeding companies we spoke to, the nucleus herd was located in Canada, and the multiplication herds located in northern France. Further breeding and back-breeding are employed in order to produce heterosis, or hybrid vigour, in high-health and disease-free conditions. The resulting parent generation is then shipped to commercial sites for on-farm breeding and production of a slaughter generation. This highly regulated and controlled system is designed to ensure that circulating pigs do not constitute a disease threat to the host farm herd.

In this sense, modern pig farming could be said to be exemplary of biosecurity using biopolitical techniques, securing the circulations that increase surplus value. The physical separation of high health nucleus units and multipliers, the programmed breeding, the continual efforts to improve genetics, the calculated circulation of pigs and semen, between sites and commercial settings, all add to the sense that disease is increasingly managed out of the system through a highly technical and modern approach to farming. Biosecurity and disease freedom are frequent claims of this system of circulation. An image that we were not able to reproduce but used by a genetics company involves an image of a pig being gifted via the cupped hands of a white-sleeved (lab-coated) scientist. The scene is under-lit, with beams of light streaming through the fingers of the scientist and illumi-nating the spotless pig. The picture is inscribed with the term 'bio-secure' and we can only surmise that the missing sub-text would be an immaculate conception. It and the other images used by genetics companies seem to confirm the edge that lab scientists have over local vets and farmers. This is not just pasteurisa-tion, it is a reduction of pig farming to pure genetics – a tendency that has been boosted more recently by the promise of gene editing and the production of dis-ease resistant pigs.

And yet, the story is not quite that straightforward. These disease-free and programmed pigs soon change complexion once we follow the process on to farms. We will detail these changes by focusing on the 'translations' that are involved in making circulating pigs into pigs that can survive on a farm.

Translations

We use the term translation to draw attention to the ways in which pigs, words, techniques and even genes are formatted in practice (Callon, 1986)[5]. Methodologically, the process involves ethnographically interrogating a putative network, like 'bio-secure farming' and 'disease-free systems', by paying attention to how these networks are actively translated in practice. So asking farmers, breeders, vets and microbiologists what it takes to be 'disease-free' and 'bio-secure' opens up a series of qualifications, or attempts to reconfigure those terms, through discussions and demonstrations of what they involve *in practice*. We will introduce these as three significant and sequential translations: first 'from disease-free to pathogen-free'; second 'from pathogen free to pathogen management'; and finally, 'from pathogen management to immuno-preparedness'.

Translation 1: From Disease-Free to Pathogen-Free

On the farm and talking to vets, we were keen to find out how the circulation of pigs and pig materials worked, and whether it was understood as being low risk and disease free. The latter, it turns out, is far from what it might seem. On the farm, there is no such thing as a disease-free pig. Instead there are attempts to assure buyers that a specific animal is clean with respect to certain specific pathogens:

> Everybody talks about being disease-free which is a nonsense. Then another word that people use as an industry is we are high health, and that's a joke. (Industry Veterinarian)

Noting that 'the only safe pig is a dead pig', pigs are not, this veterinarian points out, ever disease-free, or high health, they are instead living organisms, replete with microorganisms and the co-dependencies folded into any living organism, and more or less subject to the reproductive and production pressures of an industry under economic duress. The result is that disease-free is soon translated to a more tractable and manageable target. Here the aim is not to make disease-free animals but to make sure that certain specific pathogens are not being brought on to the farm. This involves veterinary practitioners monitoring specific pathogens and 'selling' or certifying pathogen-free, not disease-free, pigs:

> I sell specific pathogen-free animals, so I can tell you which pathogens those, my gilts or my boars are negative for... it doesn't mean that I'm selling you healthy pigs. It doesn't mean that I'm selling you disease-free pigs and then it comes down to like swine dysentery. Okay, there's a *big* pathogen... I don't sell you dysentery free pigs, I sell you brachyspira hyodysenteriae free pigs, which is the causal agent of swine dysentery. (Industry Veterinarian – emphasis added)

So disease is, in practice and for the purposes of moving pigs around, translated to the more practical, measurable and certifiable issue of specific pathogen absence. This is a reduction and economisation. Pathogens are ordered as to their importance. There are, in this excerpt, 'big pathogens' worthy of comment and action. As ethnographers, we are interested not just in *what* is said in our conversations with specialists and farmers – we are also interested in *how* things are framed and understood. The sense that some pathogens are *big*, and, by extension, there are smaller pathogens that can circulate more freely with the pigs, is a matter of analytical interest. Big here is a matter of emphasis, suggesting a microorganism that may be notable in terms of its effects and/or a life form that is difficult to eradicate or manage. Within government offices, where the focus is on trade, a similar distinction is made between exotic disease pathogens which are of most concern, including pathogens like foot and mouth viruses which are periodically eradicated, and those considered 'less dramatic' (Simmons, 2012: 349). The less dramatic include endemic diseases and minor parasitic infestations that are seldom notifiable, a legal inscription of the distinction between diseases to worry about and those that may be hazardous but not cause for concern to neighbouring farms or to the national herd, to freedom for export and so on.

There are also microbes and related diseases that are less easily ordered, and that complicate the story in other ways. Here we limit ourselves to two forms of ordering problem, one generated by pig-industry priorities and sensitivities, the other from microbiological interrelations. First, consideration of the dramatic extent of pathogens is largely framed by the pig industry and vets in ways that emphasise production rather than health *per se*. Swine flu, for example, is endemic in pig populations in many parts of the world (Irvine and Brown, 2009), with a range of clinical presentations. The infection in pigs is often subclinical and asymptomatic, which means that swine flu can be inconsequential for productivity even while pigs shed large amounts of virus. As a result, and despite a potential and even pandemic human health threat, swine flu tends *not* to be monitored systematically within the pig industry (Smith et al., 2009).

Second, in cases where pathogens interact, the distinction between mundane and consequential infection also starts to be blurred. For example, *Mycoplasma pneumoniae* is the microbial agent of Porcine Enzootic Pneumonia. It is internationally endemic in pig herds, and known to have a mild effect on pig health when animal husbandry is practised sufficiently well and when it is not found circulating with other known diseases such as PRRS (Porcine Reproductive & Respiratory Syndrome). The latter, a viral condition, sometimes known as 'blue ear' and which emerged in the 1980s, leads to a reduction in available macrophage and so reduces effective immune response in host animals. Co-infection can result in severe pneumonia, making *M. Pneumoniae* bacterium, despite its lack of notifiable status and relative impotency, an economically important disease. For this reason, and given that PRRS is difficult to control, many breeding companies and suppliers of incoming stock to farms supply pigs that are *M. Pneumoniae*-free. Controlling these bacteria is an efficient means of reducing the effects of immunosuppression in pigs that are susceptible to the PRRS virus.

Our point in detailing the swine flu and pneumonia examples is to emphasise that viruses and bacteria are categorically slippery. They confound Koch's postulates by not necessarily being pathogenic in pigs while having the potential to contribute to a dangerous disease through their microbial interactions and as a result of zoonotic transfers. In cases such as this, the dramatic ordering of pathogens becomes less straightforward, being biologically and socially complex. In practice, then, biosecurity involves a translation of disease-free into freedom from specific pathogens. Mundane (small) pathogens or, more controversially, pathogens for which the industry would rather not be called to account, are generally not included in the disease-free category. Or, if they are, it is because they form a manageable component of intercurrent conditions. This selectivity is largely based on risk to farm productivity rather than to safe life *per se*. But the translations do not stop there. A second translation is perhaps even more surprising. For in some cases it makes sense for pigs to mix with pathogens, even 'big' pathogens or those associated with important diseases, in order for them to be healthy.

Translation 2: From Pathogen-Free to Pathogen Management

In practice, things tend to be more complicated than keeping named pathogens *out*. Indeed, incoming pigs born in the rarefied and sanitised spaces of bio-secure units cannot be too naïve, immunologically speaking, in relation to their new environment, or they will soon succumb to disease themselves. So farmers and vets must seek a reasonable pathogenic match between these animals and the existing herd:

> The other side of the coin you've got is you don't necessarily want to get rid of some pathogens because if you were (…) *Mycoplasma hyopneumoniae* positive and I'm negative selling you gilts, so I have to get my gilts immunised before they get to your farm otherwise they're going to get sick, but other people want them to be negative. So that's why I use vaccines to try and ensure that my sows will live in your environment. (Industry Veterinarian)

There are two things to note here. First, being positive for a bacterium is not to say things are in a bad way on the farm. The bacterium may simply be present in pigs that have a degree of immunity to the associated disease as a result of being challenged early on in life or through maternally conferred immunity. Second, incoming pigs will need a matching pathogen exposure and so have immuno-competence roughly equivalent to the environment into which they are being transferred (or born). In short, health is not once and for all, it is actively *patched* together in ways that are *situation* specific (on the importance of patchwork in social science see Law, 2009). Even in assured circulations, this specificity matters. Indeed, circulations are not all that smooth. This is not simply a matter of sorting out bad and good circulations. To make healthy lives requires attending to specificities and the modulations that can occur between goods and bads as new constellations of lives are patched together. And, if specificity or difference matters, then circulation may not be the right term. Geographically speaking we would call this displacement a movement rather than a circulation. Crucially, these movements, or the taking into account of specificities, requires adjustments and skill by vets and farmers. Things, including pigs and their genetics, do not just flow or cascade down a breeding pyramid, they stutter and stumble and require patching together in order to move.

'Acclimatisation' is a term used in the pig sector to capture this process of patching together animal health when two sub-populations (on and off farm) meet. Vaccines are used to inoculate young pigs for some of the more dangerous diseases, but a common practice is also to acclimatise incoming pigs (either born on or transported to the farm) by exposing them to the full mix of on-farm pathogens. In stark contrast to the hyper-purity and immaculate conception of the breeding company publicity, and in a manner that reminds us of the sods of earth that Coppin mentioned, once pigs arrive or are born on to commercial premises they are routinely covered in muck:

> ... our approach has always been [to quarantine a recently arrived pig so that it] doesn't come into contact with any other pigs for at least 2 weeks post-delivery, because it takes that long for the immune system to get back to normal after the stress of moving it, and then we'd recommend that the pigs were vaccinated, and then introduce muck, from the recipient herd, whether it be weaner muck, some farms put sows in the same pens or in the next door pens, but normally what happens, muck is more effective, you gradually acclimatise them over another 4 weeks, before they're put actually into the herd. (General Manager, pig breeding company)

Methods of acclimatisation vary depending on the status of the herd, the presence or otherwise of seronegative sub-populations on site, the gestation of the disease, the built infrastructure of the farm, the bedding materials and so on. But the general point is that even in the most hi-tech versions of pig production, the use of muck provides a cost-effective and lasting means of generating attenuated forms of illness, which can then allow pigs to develop immunity to potentially costly diseases. Staging such exchanges has its own costs and dangers for the farm, as immune responses are themselves sometimes regarded as sub-optimal in terms of productivity and can produce related disorders. Nevertheless, far from being pathogen-free, pig farming is often marked by a concerted effort to generate host–microbe interactions. This not only protects pigs that are new to the farm, but also prevents these immunologically naïve animals forming a seronegative subpopulation and in turn providing the catalyst for reinvigorated microbial circulation. It also means that as replacement gilts give birth, then their piglets will have developed the necessary immunity to the farm. All of this is captured by what farmers call 'feedback':

> [...] with the gilts, we do feed back, so when the gilts come on, in theory they're into a sort of isolation unit but they get their first lot of vaccines, so they have a Blue Ear vaccine which covers for them for one shot I suppose life in effect, and then we give them the parvovirus, well avian parvo which is a combined one, but then we do feed back some of the cleansing and a few piglets, whatever, from the farrowing houses back to them, so almost as the vet would say, bug them up before they come into the herd, so in effect they pick up all the bugs that are on the herd, and he'll say 'if you've got any sort of scouring in the piglets or in the weaner house, then try and scrape up some of that into a bucket or whatever, or paper bag or something, and then give them that as well. (Pig farmer, Somerset (indoor standards)

So farmers acclimatise pigs, and vets match clients and pathogens to create a stable immunological environment, a result that can only be managed through an appreciation of the conditions on the farm. This is pathogen management as immunology management:

> I think biosecurity probably should be more considered as immunology management and what you're trying to do is to stabilise the farm immunologically and not have naive animals coming into a positive pathogen herd and likewise positive pathogen animals coming into a naive herd. It may well be that there are some pathogens where ... congenital tremor virus, which we don't know what it is, causes horrendous problems in some farms. A small farm for instance, can go negative, so then they buy pigs from a breeding company which is naturally positive and then we have devastation through that. So then you've got to match clients. (Industry Veterarian)

Crucially, as this interview extract also suggests, immunological management is never a once and for all. Farms can go negative for some diseases that are otherwise endemic. There are also the continuous shifts and drifts in microbial strains as well as new interactions that result from constantly changing microbial environments. PRRS, for example, arose in the 1980s partly as a result of pathogen management. Selectively removing microorganisms sets up conditions of possibility for new microbial and biotic interplays, and the result can make, as we have seen, the otherwise less consequential microbes, like *M. Pneumoniae*, more of a concern (or in terms we used earlier, it increases their pathogenicity). So microbes and immune responses are not controlled, so much as they are managed. Things can go from small to big as a result of management and shifts in micro-ecologies. And this is management in the sense of a continuous, ongoing coping and responsive adjustment. It is an experimental process, we would suggest, with the aim of producing *enough* health through pathogen matching and patching. And yet, what is most remarkable about pigs in muck and the dynamics of disease ecologies is not so much the careful regulation of circulation, but the re-emergence of movement and the spatial openness of communing with microbial life. We return to this openness in the next section of this chapter.

For now, we have traced two translations of the disease-free concept. We have moved, as a result, from a dichotomous sense of health *or* disease to the idea of health as a quantity, something that Georges Canguilhem's (1991 (1966)) famous critique of the normal and the pathological might have anticipated. In short, and in practice, we have moved to the practical process of producing 'enough' health through a patchwork of materials and practices. This may not be altogether surprising, and certainly other researchers who have looked at the use of antibiotics in the pig sector confirm that disease and health are practised not as an either/or but as a continuum. Coyne et al. (2014) confirm that for pig farmers and vets in the North West of the UK:

> Disease was not considered as a static state in which a pig is either infected or not, it was considered as a dynamic state in which different diseases interact to form a 'stew pot of disease' and where disease may be sporadic, sub-clinical or a persistent problem (Coyne et al., 2014: 4).

The final translation takes this one step further, for as soon as health becomes a patchwork of practices, then the focus can shift again. 'Pathogens' start to lose their stability, their timelessness, and immunity starts to become the object of attention. To be clear, immunity and immune systems are hardly innocent 'objects of belief' (Haraway, 1991: 204) and may shore up rather than unsettle boundaries between the normal and the pathological, self and other, as well as reproducing militaristic senses of fighting disease (Waldby, 1996). But, as we will show, the diminishing returns of pathogen management can seed and in turn provide glimpses of other ways of configuring health – ways that involve, we will argue, contingent, situated and practical engagements.

Translation 3: From Pathogen Management to Immuno-Preparedness

As we have already noted, biosecurity can in practice be translated to 'immunology management'. Here vaccination and acclimatisation are matched to improvements in husbandry. The latter involves attempts to ensure that pigs are healthy, and immune systems are correspondingly 'strong':

> What you're trying to do is to build a disease-resistance into your livestock and to give their immunity every chance to strengthen by making sure husbandry's absolutely right. (Mixed farmer, Wiltshire, organics)

Husbandry includes matters as diverse as airflow, welfare and stress management, as well as developing means for pigs to be challenged by pathogens. It is, as Roe et al. (2011) have demonstrated, a multiple and often tacit set of skills and practices that stock people learn by doing. Such knowledge practices are situated and arguably under threat within an increasingly automated and regulated system. But more than a matter for tacit knowledge and sensitised 'subjectivities' alone, there are also important shifts to note here in the 'objects' of concern. First, there is a shift of attention from pathogens to health, a relational move wherein pathogens and microbes figure but are no longer determinate (i.e. pathogenicity is somewhat less decided). Second, there is a move from delimiting boundaries between inside (secure) and outside (insecure) to a focus upon the *quality* of embodied relations, or the *ways* in which inside and outside are joined together. Immunity in this sense becomes a matter of preparedness in terms of both known and unknown challenges to health, making sure that pigs were experienced enough to withstand inevitable challenges:

> Just get them bugged up, so expose them to all the bugs that are on the farm, and then that in theory should obviously boost their [immune system], they'll have a big sort of flush of whatever and then hopefully in the farrow they'll pass on their immunity to the piglets. (Pig farmer, Somerset, indoor standards)

So, in contrast to the stark divide of having a specified pathogen either present or absent, here the farmer encourages the communing and subsequent alteration of 'pathogen' and 'animal'. Pig farming, in this sense, is more about hands-on, generative and situated knowledge than solely the implementation of industry standards and norms. This is different to the version of bio-communicability that often accompanies accounts of global health threats and food insecurity. There, bio-communicable suggests a world of contagion; a touching together that is regarded as symptomatic of weakness and inevitable disaster. Here, the communing of animals and pathogens is potentially life-affirming, where the sharing of the 'munus' is more akin to the munificence or gift of community. There is a conferring of health by, and not in spite of, an organism's environment. Community and immunity make one another – they are folded together rather than in opposition (Esposito, 2008, 2011). The heterogeneous practices of animal husbandry are in effect methods of producing health through processes of learning and embodied memory, creating what we would characterise as both immunity to known challenges but also a more broadly defined immuno-preparedness, where the object of concern is less defined:

> I think so much livestock production is about, 'Let's annihilate all the bacteria.' And often… you really need…to nurture those bacteria which give you a healthy system, and that's where natural immunity comes from. (Mixed farmer, Devon, outdoor system)

Immunity then is not located within the body of an animal, or at the farm gate, it is the emerging product of many relationships. It is on the farm, embodied and memorised within a herd, conferred maternally and facilitated through husbandry. It is also about managing the units on the farm, making adjustments to minimise stress:

> If you're constantly mixing, you're stressing them, it makes their immune system lower as I understand so they're more susceptible to clostridial diseases, coli's. *Streptococcal meningitis* is really, really triggered by stress, it can be triggered by sudden changes in temperature or aggravation, anything that puts them out of kilter, that's probably the biggest one within the pig industry, meningitis. (Pig farm manager, Devon, outdoor freedom foods.)

And the results may well be surprising. This farm had re-introduced nesting materials, practised free-range farming and focused on animal welfare at the same time as adopting all-in-all out and closed herd management. Even so, productivity had not been compromised according to the manager:

> ...behind there is a little pot hole at the end of each pen so inside there, there's a heater and a straw box so the piglets can go in there to get away from Mum and it's lovely and cosy in there.

> This is the big selling point for this farm, we're getting an intensive pig unit's production, we're in the top 10% performing herd in the country, on a system that your grandad would have used, you know? (Pig farm manager, Devon, outdoor freedom foods.)

There is a risk of nostalgia and romance here of course. But the key point is that health is borne from what Law (2004b), following Kwa (2002), might call a baroque rather than romantic conception of relational space. In other words, there is no total system here, or a whole that can be invoked as something towards which we must reach or return. Rather, the patching together we have detailed is non-coherent, 'materially heterogeneous, specific, and sensuous' (Law, 2004b: 13).

To summarise, modern pig farming is characterised by movements as well as circulations; contact as well as exchanges; and communication as well as commerce. To be sure their relative ascendency will differ depending on the farm. But our point is that they are not so much alternatives as forms of displacement that co-exist. The network of pig circulations cannot, it seems, hold, without the movements and communications that make healthy pigs.

This co-existence of forms of displacement produces and is produced by a tension between disease diagrams. There is the familiar inside/outside binary that underlines contamination versions of disease. All that biosecurity and disease-freedom borne of fears over trade-limiting diseases gives birth to immaculately conceived pigs. But it is matched, or co-exists with other diagrams that share more with security and the interventions involved in vaccines, inoculations, breeding, genetics, acclimatising and husbandry. These are not, though, applied in a uniform manner. There are particularities and specificities.

In the terms of the depasteurisation of England, we might note that the 'edge' that the scientist and 'his' lab gets over local veterinarians and farmers is starting to look somewhat less sharp. The 'dance of agency' (Pickering, 2008) involves not only lab scientists, vets and farmers, but also pigs, buildings, microbes, markets and so on. Here immunity starts to change the contours of security, it shifts the objects of attention and the cast of characters who might be involved in making life healthy. But we also need to move carefully here, for we would not be the first to proclaim that immunity and health are the best means of disease prevention. These terms may not be as progressive as they first appear. In the final section, we look at immunity within social theory and the attention to health in a little more detail. Our argument is that the practices we have witnessed must not only change the means of achieving good health, they also need to challenge the ends of good life.

Immunity, Attention and More-than-Human Responses

Immunity is a tricky term with which to work. Initially referring to legal exemptions (from taxes, or public service), it tends to suggest some kind of absolution, impunity even (Esposito, 2011). To be immune from prosecution is to be made impervious to the norms of jurisdiction. Immunity in this sense tends to refer to an exceptional state, a rising above or beyond the community. The ability of some to avoid a second bout of an infectious disease was clearly related to this sense of exemption. But as the term gained greater traction in the mid- to late twentieth century within medicine, the notion of an exempt self was displaced by a networked body (Anderson and MacKay, 2014). Rather than being a straightforward science of the self, or an account of how individual selves were somehow abstracted from their milieu or community, immunity started to become regarded as a situated and shared space.

Within social science, and particularly following the emergence of Auto-Immune Deficiency Syndrome (AIDS), immunity has tended to be regarded with some suspicion. It is a term that is easily hitched to social meanings and currencies, including most notably militarist notions of defence and corporatist versions of flexibility. Emily Martin's ethnographic study of the science and culture of immunity, for example, highlighted not only the persistence of martial metaphors within scientific and popular discourse on immunity, but also the rise of complex systems thinking which allied immunity to archetypically post-Fordist tropes of flexibilisation and just-in-time production (Martin, 1995). The point is of course that our understanding of immunity is hardly innocent, both in terms of the metaphors that are generated in its representation and in terms of the work that 'the' science can then do in terms of providing legitimacy and authority to social formations.

These warnings are important, particularly as we do not want to be misread here and elsewhere in the book regarding a shift to immunity that is either exclusive or indeed all-encompassing and therefore part and parcel of a liberal, ethicised state of being. In the context of this chapter, we do not read immunity management as either another term for keeping things out, simply marking a shift in boundaries from the farm to the body. Nor do we see it as a regressive move for pig farmers who are suddenly tasked with the endless job of maintaining a constant watch on behalf of global health. In a more positive vein, we see immunity as presenting at least three opportunities for re-diagramming pathological lives:

1. First, immunity acts to trouble continuously any reductive account of medicine and disease causation (exemplified in germ theory) through a focus on the variations in disease causation (Anderson and MacKay, 2014: 4) or what we have called disease situations. Immunity in this sense puts paid to any account of disease that equates pathogen presence with disease. Clearly, and as our accounts of pig farming have suggested, presence of a microbe is, on its own, often insufficient for a disease to develop.
2. Second, immunity is a collective property, not one that resides within individual selves. Incoming pigs will, as we noted, not only potentially add new microbes to a farm's microbiome, they will also potentially disrupt the herd immunity that exists on a farm. Herd immunity, which is as applicable to people as it is to pigs, is conferred when a reliable proportion of any given population is capable of reducing the effectiveness of a microbe's replication and onward transmission (its reproduction rate or R_0). To put it simply, even a well-vaccinated and so largely immune individual will be more prone to a disease if he or she is placed within a largely unvaccinated community. Conversely, a relatively unprotected individual will be safer in a largely vaccinated community. In this sense, as the novelist Eula Biss notes, 'we are protected not so much by our own skin, but what is beyond it' (Biss, 2014: 25).
3. Finally, and to take this displacement of selves even further, immunity is not only a shared property with others like us, but it also accentuates the transspecies natures of any organism. Gut microbiomes are replenished by acclimatisation of pigs on a farm, helping to reduce scour in piglets. 'We' are networks rather than self-contained beings. Microbes not only provide the conditions for living, they also induce gene expression and, indeed, provide the raw material for what we call our own immune responses. So much so that the topological character of immune cells are regarded as nothing more and nothing less than adapted viruses.

So immune selves do not so much keep disease at bay as perform pathological lives through engaging in an ordering process that involves, or is contiguous with, all manner of others, not all of whom are necessarily benign. This troubling of any hard and fast division between self and non-self is the condition for an immune system that can easily flip from being constructive to destructive, in the form of autoimmune diseases (Anderson and MacKay, 2014). Indeed, it is this 'trickster' quality of immunity (Haraway, 1993) that seems, at first blush, to offer something other than a continual circulation of metaphors and borrowings between science and culture. For Haraway, this is a world not only of cultural circulations but also of ongoing and always surprising encodings that trouble any neat tales of causation:

> So, while the late twentieth-century immune system, for example, is a construct of an elaborate apparatus of bodily production, neither the immune system nor any other of biology's world-changing bodies – like a virus or an ecosystem – is a ghostly fantasy. Coyote is not a ghost, merely a protean trickster (Haraway, 1993: 298).

As Haraway goes on to suggest, as immune systems become scientific entities, so the object of interest shifts from germs to networks. In Whiteheadian fashion, the complexities of research work and the things that animate that science seem to force a shift, at first subtle, and later overwhelming, from reductionist accounts of disease:

> Unable to police the same boundaries separating insiders and outsiders, the world of biomedical research will never be the same again (Haraway, 1993: 324).

This network body draws us back to our interest in disease situations, to topology and to what we briefly referred to as borderlands. Importantly, immune systems science does not replace a topography of exclusion with a world of flux, cross-border solidarity or a de-differentiated world. Indeed:

> The pleasures promised here are not those libertarian masculinist fantasmics of the infinitely regressive practice of boundary violation and the accompanying frisson of brotherhood, but just maybe the pleasure of regeneration in less deadly, chiasmatic *borderlands* (Haraway, 1993: 306, emphasis added).

While immune systems research may underline the fact that selves and identities are not simply threatened but also *earned* through interactions with biological difference (Napier, 2012), it also starts to shift the spatial coordinates of pathological lives. Returning to our fieldwork with pigs, microbial life becomes central to making health possible, a necessary component in patching together livestock and health. Here the viruses and bacteria act less like single-minded attackers and rather stand 'somewhere at the borders of "self" and "non-self" and act as conditioners and definers "of a body's boundaries"' (Napier, 2012: 126). They inhabit a borderlands, a zone of mixed orders where inside and outside are displaced by a mix of spatio-temporal logics that can qualify as well as disqualify life.

This immunological move that troubles self and non-self lies at the heart of much feminist, social and political theory. But this sense of conditioning, embedded within immunity practices and in field terms that we have witnessed like acclimatisation, again draws us close to and allows us to expand on the concerns of the political and legal theorist Esposito (2008, 2011).

For Esposito, immunity is distinct from exclusion, where 'life is preserved inside an order that excludes its free development because it is retained within the negative threshold defined by its opposite' (Esposito, 2011: 10). Exclusion is, then, a self-preservation that risks the reduction of life to the same, to the merely 'us' (see Chapter 4 for an example of this self-destruction through exclusion). As Esposito and many others have pointed out, the result of such preservation can be a hypertrophic security, a self-protective syndrome where attempts to exclude difference are ultimately self-destructive. The result contributes to what Campbell (2011) and others see as a somewhat inevitable reduction of life to its bare bones, to its merest of qualities, and what Agamben regards as a *making survive* (Agamben, 1999: 155). Immunity, by way of contrast, emerges as a 'a non-excluding relation with the common opposite' (Esposito, 2011: 17). But, to be clear, non-excluding does not mean inclusive. For Esposito, Haraway and for the farmers, vets and breeders we have spent time with, this is not a simple and rather romantic ideal of no-borders. Biological difference still matters and should not be eradicated through exclusion, nor, at the other extreme, denied through unconditional hospitality. Rather there is a continuously changing world, a borderland of folded relations, within which health is made and remade. And, here, it is the memory of immunity that seems key. This alerts us to the Bergsonian sense of life as *interval* between action and reaction (Bergson, 1988). This is the responsiveness of life, perhaps (Derrida, 2003). For Bergson the matter of memory takes us very close to this form of life that does not move through a self, and thereby can resist a declension of biopolitics to a power over life, or a mastery of the world. As we understand it, memory is inward inflexion, a mattering that is taken up by Deleuze in his notion of the virtual (Deleuze, 1991; Lash, 2006). There are spatio-temporal resources here for discussing the immanence of immunity, not in terms that reduce it to a flat world, of inclusion, but that work with the inflexions, the foldings together and the memory-making that make life possible. The result is to make microbes seem not so much as outsiders but as co-constituting the borderlands that make any life possible

Our final question relates to how this shift of attention, from borders to borderlands, challenges or indeed re-asserts a politics of life? What in other words does the depasteurisation of England look like in practice? There are two common responses to elevating immunity as a means to health. The first concerns a paradox relating to the new mix of a rekindled interest in health as a constitutional or configurational quality (something that we can intervene in) being coupled to a complex system that is almost by definition beyond control. The second relates to the appropriation of immunity within security discourses. We take each in turn before returning to the issue of the means and ends of a politics of life.

First, returning to Emily Martin's critical appraisal of the ideologies that attach to immunity discourse, there is a warning that any shift from the simplicities of keeping disease out to an understanding of health and immunity as a complex system is freighted with consequences in terms of what this means for an active role in making health. Immunity in many ways involves both a re-centring of agency for health (health becomes a project of the self) at the same time as a displacement of the self such that health is conferred in a shared space over which the agent has relatively little control. The result is a pernicious and in many ways self-defeating responsibilisation of individuals for their (or their animals') health. Martin rightly highlights the overwhelming task of becoming responsible for one's own health at the same time as realising the distributed nature of that project. This is the 'paradox of feeling responsible for everything and powerless at the same time, a kind of empowered powerlessness' (Martin, 1995: 122). The result can, of course, be an injunction for a constant vigilance that in itself produces the auto-immune responses of stress-related disease.

Second, it is notable that the complex un-decidability trickster of immunity has not so much undermined martial discourses of defence and exclusion as provided new scope for a securitisation of life. For Haraway, writing in the early 1990s at the height of a progressive response to the horrors of AIDS and the reactionary politics it seemed to unleash, and in the aftermath of a decade of Cold War posturing that involved defence systems modelled on immunity, the new science of distributed immunity offered promise of better things to come. Intrigued and intoxicated with the promise of this trickster called the immune system, Haraway wrote:

> The concatenation of internal recognitions and responses would go on indefinitely, in a series of interior mirrorings of sites on immunoglobulin molecules, such that the immune system would always be in a state of dynamic internal responding. It would never be passive, 'at rest', awaiting an activating stimulus from a hostile outside. In a sense, there could be no exterior antigenic structure, no 'invader', that the immune system had not already 'seen' and mirrored internally. Replaced by subtle plays of partially mirrored readings and responses, self and other lose their rationalistic oppositional quality. A radical conception of connection emerges unexpectedly at the core of the defended self. Nothing in the model prevents therapeutic action, but the entities in the drama have different kinds of interfaces with the world. The therapeutic logics are unlikely to be etched into living flesh in patterns of DARPA's (US Defense Advanced Research Projects Agency) latest high-tech tanks and smart missiles (Haraway, 1993: 323).

The message seems clear, 'partially mirrored readings and responses' replace any firm distinctions between inside and outside. In our example, it is the *qualities of the* relations between animals, microbes and people that are key and any tendency to reduce those relations to universals like disease-free or bio-secure (meaning sanitised) agriculture can be part of the problem. They risk damaging the heterogeneous immunity that enables this assemblage to meet new threats.

And yet, Haraway may have been too quick to rule out the appropriation of this reconfigured immunity in agencies like DARPA and within biosecurity more generally. Indeed, the shift from hierarchical bodies to a networked body of 'amazing complexity and specificity' may have re-galvanised security thinking and action. Scientific and ontological complexity and specificity have not, it seemed in the following two decades, resulted in a more modest politics or the predicted humility that many in science studies may well have, implicitly at least, believed. Indeed, as Melinda Cooper's work amongst others has detailed, indeterminacy and complexity are as likely to result in redoubled efforts to secure futures as well as usher in shifts in the modes of addressing those futures. Preparedness and other anticipatory modalities in this case become associated with constant vigilance within a state of continuous becoming:

> From climate change, to emergent disease and bio-safety, there is a moral injunction to anticipate as an act in which life, death, identity and prosperity are at stake personally and collectively. Anticipation calls for a heralding of the emergent 'almost' (Adams et al., 2009: 254).

Preparedness, for example, has its own history and is certainly non-innocent (Collier and Lakoff, 2008a; Cooper, 2008). In distinction to prevention and precaution, preparedness 'does not aim to stop a future event happening. Rather intervention aims to stop the effects of an event disrupting the circulations and interdependencies that make up a valued life' (Anderson, 2010: 791). As such, and as we suggested in Chapter 2, it can signal another instance of hyper-trophic security, a vigilance for as yet undetermined threats. This is the making present of emergent dangers.

Attending to immunity may, in this case, be read as another instance of this expansion of the sphere of influence of security. This is, for some authors, the grasping quality of the security diagram. As we noted in Chapter 2, Lentzos and Rose are clear that Foucault's specification of security and liberalism was indeed centrifugal and therefore likely to both encounter *and* encompass difference as a means 'to manage or regulate that complex reality towards desired ends' (Lentzos and Rose, 2009: 232).

So, on this view, biosecurity as disease-free agriculture unsurprisingly requires the supplement of contingent variation in order to succeed. Security requires a surfeit of life, wherein even the most immaculate and international of living spaces needs faeces for the network to hold. Patching together has become not just the leitmotif of science and technology studies keen on undermining any grand narratives of progress. It has also become a matter of practice for all manner of actors in various sectors as they are subject to the injunctions to remain vigilant and ever watchful for the emergent 'almost'. A shift to healthy bodies in this sense can hardly be said to be unquestionably progressive given the injunctions of a new public health or animal health to become exemplary bodies and look after oneself.

We are sympathetic to these warnings, and to the work of authors like Caduff (2014), who argue that a cosmology of mutable life has produced a space for dire pandemic prophecy and for uncertainty as justification for hyper-security. And yet, the shift of focus from pathogens to pathogenicity, and from fixed objects to shifting relations, can also suggest a different opening. So how might immunity open (rather than guarantee) a different kind of politics, one that may be less prone to capture? Our point, for now, is that the embodied and visceral intra-actions of pigs, people, microbes and health may start to undermine the very programme of action or desired *ends* that is biosecurity. Working alongside pigs and bugs, the practitioners we spoke to start to trace a life that is lived at the limit of itself, on the borderlands. Here the desired end loses some its stability so that 'disease is no longer configured as extreme risk, but rather as the risk of not being able to face new risks' (Esposito, 2008: 191). The risk of not being able to face new risks requires not so much a new set of responsibilities or 'new normals' but a counter-norm of communing with others, and the openness to difference that confers immunity. The surfeit of microbial relations and the ecology of practices (Stengers, 2010b) that are involved in making health suggest to us not so much a power over life or indeed only a power of life that can be turned against itself (Chapter 4). Rather, health is always a matter to be worked out, and patched together in heterogeneous conditions (Chapters 8 and 9).

To be clear, the point of raising these practices does not suggest that immunity can stand outside the ongoing formulation and reformulation of life politics. We are aware that affirmations of proper life, even folded ones where communing with difference is emphasised, are prone to capture and codification in ways that reduce their potency and configure new norms. Nevertheless, in demonstrating that the current network or programme of action is held together by pig muck and by various acts of patching together lives that always involve more-than-human lives, we reveal both the value of immunity and the limits to a disease-free model of farming where a de-skilling of both pigs and people can undermine a capability to meet new risks. Following translations has enabled us to identify those aspects of a network that need to be conserved or given value, not simply to maintain the biosecurity network (though covering pigs in muck clearly does this), but to challenge its tendency to homogenise practice. The label disease-free tends to downplay the required situated and dynamic knowledge that pigs, people and microbes fold together on the farm. In this, our practitioners are not *responsible* for biosecurity, but *responsive* to living complexities and make valuable contributions towards making life safe. In other words, making life safe requires a pig sector where pigs as trans-species, living, excreting, moving, communicating and co-muning animals are factored into the debate. Likewise, it requires us to recognise the importance of knowledge practices, not as a means to simply redistribute disease costs and responsibilities and generate neo-liberal subjectivities, but as speculative contributions to a vital ecology of practices.

Conclusions

Depasteurising England requires all manner of actors, including biomedical sciences, animal sciences and farming experts, vets and others, local and otherwise. Assembling that expertise is a key task and one that should resist any further erosion of those skills and practices, as well as the isolation of key actors as solely responsible for health. Immunity is a shared space, one that requires collective and social action that is responsive, attendant to the biographies, constitutions and therefore situations of health and disease. There is, we would argue (and following Coppin), nothing inevitable about the current evolution of pig farming. Indeed, in terms of the expert practices we have witnessed, there is a more than enough indication that an alternative history and geography of livestock farming is possible. Presenting such an account would change the current emergency of infectious disease to one of living responsive, pathological lives.

Endnotes

1   Interestingly, the plans were rejected by the Environment Agency in 2015 on the grounds that the mega farm posed risks to human health and human rights. The site was next to a women's prison, and one of the arguments was that the incarcerated women could not move away from the confined pigs. The traffic between incarcerated bodies of various kinds was an interesting sub-text to the campaign.

2   In 2013 there were roughly 450,000 breeding sows in the UK, producing around 9 million pigs for slaughter per year. Around 60% of these sows were housed indoors, with the remainder outdoor, but nearly all their offspring were finished indoors (98% of the 9M total) (BPEX, 2013). At the 2013 census, there were 4.4 million pigs in the UK (DEFRA, 2014).

3   We should note that there are differences between retailers in the UK, with some providing forms of market organization that are less prone than others to the effects we detail here. While the bulk of sales are characterized by instability, differences at the higher and lower ends of the retail sector are interesting. See Bowman et al. (2012).

4   The highest category mapped in 2006 was > 80,000 pigs per UK census area. Devon and Somerset were in this category along with more traditional pig areas of North and East Yorkshire, Lincolnshire, Norfolk and Suffolk (Defra, 2006).

5   Callon's (1986) use of the term broadly referred to the ways in which organisms, materials and people are all subject to transformations as they are enrolled into specific socio-technical arrangements or programmes of action.

References

Adams, V., Murphy, M. & Clarke, A.E. 2009. Anticipation: Technoscience, life, affect, temporality. *Subjectivity*, 28, 246–265.
Agamben, G. 1999. *Remnants of Auschwitz*. New York: Zone Books.

Anderson, B. 2010. Preemption, precaution, preparedness: Anticipatory action and future geographies. *Progress in Human Geography*, 34, 777–798.

Anderson, W. & Mackay, I.R. 2014. *Intolerant Bodies: A Short History of Autoimmunity*. Baltimore, MD: Johns Hopkins University Press.

Bergson, H. 1988. *Matter and Memory*. New York: Zone Books.

Biss, E. 2014. *On Immunity: An Inoculation*. London: Fitzcarraldo Editions.

Bowman, A., Froud, J., Sukhdev, J., Law, J., Leaver, A. & Williams, K. 2012. Bringing home the bacon: From trader mentalities to industrial policy. *CRESC Public Interest Report* http://www.cresc.ac.uk/publications/bringing-home-the-bacon-from-trader-mentalities-to-industrial-policy

BPEX 2013. *British Pig Executive Website* [Online] http://www.bpex.org.uk/prices-facts-figures/industry-structure/UKpigbreedingherd.aspx

Caduff, C. 2014. Pandemic prophecy: Or, how to have faith in reason. *Current Anthropology*, 55, 296–305.

Callon, M. 1986. Some elements of a sociology of translation: Domestication of the scallops and the fishermen of St Brieuc Bay. *In*: Law, J. (ed.), *Power, Action and Belief*. London: Routledge and Kegan Paul.

Campbell, T.C. 2011. *Improper Life: Technology and Biopolitics from Heidegger to Agamben*. Minneapolis: University of Minnesota Press.

Canguilhem, G. 1991 (1966). *The Normal and the Pathological*. New York: Zone Books.

Collier, S.J. & Lakoff, A. 2008a. Distributed preparedness: The spatial logic of domestic security in the United States. *Environment and Planning D: Society and Space*, 26: 7–28.

Cooper, M. 2008. *Life as Surplus: Biotechnology and Capitalism in the Neoliberal Order*. Seattle: University of Washington Press.

Coppin, D. 2003. Foucauldian hog futures: The birth of the mega-hog farms. *Sociological Quarterly*, 44, 597–616.

Coppin, D. 2008. Crate and mangle: Questions of agency in confinement livestock facilities. *In*: Pickering, A. & Guzik, K. (eds), *The Mangle in Practice: Science, Society and Becoming*. Durham, NC: Duke University Press.

Coyne, L.A., Pinchbeck, G.L., Williams, N.J., Smith, R.F., Dawson, S., Pearson, R. B. & Latham, S.M. 2014. Understanding antimicrobial use and prescribing behaviours by pig veterinary surgeons and farmers: A qualitative study. *Veterinary Record*, doi: 10.1136/vr.102686.

Cussins, C. 1996. Ontological choreography: Agency through objectification in infertility clinics. *Social Studies of Science*, 26, 575–610.

DEFRA 2006. *Distribution of pigs in UK on 2 June 2005* http://archive.defra.gov.uk/foodfarm/farmanimal/diseases/vetsurveillance/reports/documents/rp6149.pdf. London, UK.

DEFRA 2014. *Farming Statistics – Livestock Populations at 1 December 2013, United Kingdom* https://http://www.gov.uk/government/uploads/system/uploads/attachment_data/file/293717/structure-dec2013-uk-19mar14.pdf. London, UK.

Deleuze, G. 1991. *Bergsonism*. New York: Zone Books.

Derrida, J. 2003. And say the animal responded. *In*: Wolfe, C. (ed.), *Zoontologies: The Question of the Animal*. Minneapolis: University of Minnesota Press.

Dunn, E.C. 2005. Standards and person-making in East Central Europe. *In*: Ong, A. & Collier, S.J. (eds), *Global Assemblages: Technology, Politics and Ethics as Anthropological Problems*. Oxford: Blackwell.

Esposito, R. 2008. *Bios: Biopolitics and Philosophy*. Minneapolis: University of Minnesota Press.

Esposito, R. 2011. *Immunitas: The Protection and Negation of Life*. Cambridge: Polity Press.

Geels, F.W. 2009. Foundational ontologies and multi-paradigm analysis, applied to the socio-technical transition from mixed farming to intensive pig husbandry (1930–1980). *Technology Analysis & Strategic Management*, 21, 805–832.

Haraway, D. 1991. *Simians, Cyborgs, and Women: The Reinvention of Nature*. London: Free Association Books.

Haraway, D. 1993. The promises of monsters: A regenerative politics for inappropriate/d others. *In*: Grossberg, L., Nelson, C. & Treichler, P. (eds), *Cultural Studies*. London: Routledge.

Holloway, L., Morris, C., Gilna, B. & Gibbs, D. 2009. Biopower, genetics and livestock breeding: (Re)constituting animal populations and heterogeneous biosocial collectivities. *Transactions of the Institute of British Geographers*, 34, 394–407.

House of Commons Food and Rural Affairs Committee 2009. *The English Pig Industry*. http://www.publications.parliament.uk/pa/cm200809/cmselect/cmenvfru/96/96.pdf

Irvine, R.M. & Brown, I.H. 2009. Novel H1N1 influenza in people: Global spread from an animal source? *The Veterinary Record*, May 9, 577–578.

Kwa, C. 2002. Romantic and baroque conceptions of complex wholes in the sciences. *In*: Law, J. & Mol, A. (eds), *Complexities: Social Studies of Knowledge Practices*. Durham and London: Duke University Press.

Lash, S. 2006. Life (vitalism). *Theory, Culture and Society*, 23, 323–329.

Latour, B. 1983. Give me a laboratory and I will raise the world. *In*: Knorr-Cetina, K.D. & Mulkay, M. (eds), *Science Observed. Perspectives on the Social Study of Science*. London: Sage.

Law, J. 2004a. *After Method: Mess in Social Science Research*. London: Routledge.

Law, J. 2004b. And if the global were small and noncoherent? Method, complexity, and the baroque. *Environment and Planning D: Society and Space*, 22, 13–26.

Law, J. 2009. Actor-network theory and material semiotics. *In*: Turner, B.S. (ed.), *The New Blackwell Companion to Social Theory*. Oxford: Blackwell.

Lentzos, F. & Rose, N. 2009. Governing insecurity: Contingency planning, protection, resilience. *Economy and Society*, 38, 230–254.

Marquer, P., Rabade, T. & Forti, R. 2014. Pig farming in the European Union: Considerable Variations from one Member State to Another. *EUROSTAT* http://ec.europa.eu/eurostat/statistics-explained/index.php/Pig_farming_sector_-_statistical_portrait_2014 Accessed 18 September 2015.

Martin, E. 1995. *Flexible Bodies: The Role of Immunity in American Culture from the Days of Polio to the Age of Aids*. Boston, MA: Beacon Press.

Midland Pig Producers Ltd. 2012. *Foston Development*. [Online] http://www.mppfoston.com/

Mitropoulos, A. 2012. *Contract & Contagion: From Biopolitics to Oikonomia*. New York: Minor Compositions.

Napier, A.D. 2012. Non-self help: How immunology might reframe the enlightenment. *Cultural Antrhopology*, 27, 122–137.

Parliamentary Office of Science and Technology, 2012. *Livestock Super Farms. POSTnote*. http://www.parliament.uk/post London: Parliamentary Office of Science and Technology.

Pickering, A. 2008. New Ontologies. *In*: Pickering, A. & Guzik, K. (eds), *The Mangle in Practice: Science, Society and Becoming*. Durham: Duke University Press.

Roe, E., Buller, H. & Bull, J. 2011. Using your eyes and ears: The performance of on-farm welfare assessment. *Animal Welfare*, 20, 69–78.

Short, K.R., Mathilde, R., Verhagen, J.H., Van Riel, D., Schrauwen, E.J.A., Van den brand, J.M.A., Manz, B., Bodewes, R. & Herfst, S. 2015. One health, multiple challenges: The inter-species transmission of influenza A virus. *One Health*, 1, 1–13.

Simmons, A. 2012. Food security and infectious animal and plant diseases: A UK government perspective. *Geographical Journal*, 178, 348–350.

Smith, G.J.D., Vijaykrishna, D., Bahl, J., Lycett, S., Worobey, M., Pybus, O., Ma, S.K., Cheung, C.L., Raghwani, J., Bhatt, S., Peiris, J.S.M., Guan, Y. & Rambaut, A. 2009. Origins and evolutionary genomics of the 2009 swine-origin H1N1 influenza A epidemic. *Nature*, 459, 1122–1126.

Stengers, I. 2010b. Including non-humans in political theory: Opening Pandora's Box? *In*: Braun, B. & Whatmore, S. (eds), *Political Matter: Technoscience, Democracy, and Public Life*. Minneapolis: University of Minnestoa Press.

Taylor, S.J. & Bogdan, R. 1984. *Introduction to Qualitative Research Methods*. US: Wiley and Sons.

Veterinary Medicines Directorate 2013. UK Veterinary Antibiotic Resistance and Sales Surveillance 2012. *UK-VARSS 2012. UK Veterinary Antibiotic Resistance and Sales Surveillance Report*. The combined report is available from http://www.vmd.defra.gov.uk London, UK.

Waldby, C. 1996. *Aids and the Body Politic: Biomedicine and Sexual Difference*. London: Routledge.

Woods, A. 2012. Rethinking the history of modern agriculture: British pig production, c. 1910–1965. *Twentieth Century British History*, 23, 165–191.

Chapter Six
Attending to Meat

Introduction

In early 2013, processed meat products, which were sold as 'beef-based' in several UK and Ireland supermarkets, were found to be adulterated with as much as 100% un- or improperly declared, criminally-introduced horsemeat, a situation that was subsequently found to be widespread within the industry. Although there was little suggestion that human health had been compromised by this adulteration, the scandal followed a series of others (food scares such as *Salmonella* in the 1980s, BSE in the 1990s, and farming crises such as Foot and Mouth Disease in the 2000s). They all drew attention to the fragility of the systems through which food is produced in the UK (and elsewhere), and prompted the need to understand how such a fundamental constituent of social life may (better) be governed.

This chapter engages with this question of governing food, asking how the diverse materials and materialities, places and traces, qualities and demands that are made of food, can be managed. Informed by an analysis of the breakdown in food governance represented by the horsemeat 'crisis' and fieldwork undertaken with the inspectors tasked with enforcing food safety with the UK's food business operators (FBOs), we argue that answers to the challenge of keeping food safe must eschew the temptations of simple solutions or silver bullets and instead operate through an appreciation of how the different facets of food are coordinated and taken care of in practice.

Pathological Lives: Disease, Space and Biopolitics, First Edition. Steve Hinchliffe, Nick Bingham, John Allen and Simon Carter.
© 2017 John Wiley & Sons, Ltd. Published 2017 by John Wiley & Sons, Ltd.

The chapter proceeds in four main sections. The first section (Mapping the Current Landscape of Food Safety) reviews how the UK post-war and particularly post-BSE experience of food production and consumption have shaped the current suite of logics and practices through which food safety is regulated. The second section (A Failure of Coordination?) outlines how the aforementioned meat adulteration event, as well as the issue of *Campylobacter* in poultry explored in Chapter 4, has recently served to expose some of the limits of that configuration. The third section (Inspection as Tending the Tensions of Food Safety) draws on fieldwork undertaken on the practices of food safety through the UK food 'chain' to explore the extent to which the technology of inspection offers a locus of the kind of regulatory expertise and skilled practice that is required to attend to the situations (and not just the sites) of food safety. Finally, the last section (Being Stretched) focuses on how current pressures on this expertise and skilled practice are risking its marginalisation and indeed loss from the landscape of food safety regulation, itself a situation with potentially seriously consequences.

Mapping the Current Landscape of Food Safety

In their thorough review of the development of the 'New regulation and governance of food' in the UK, Marsden et al. (2009) helpfully identify three key stages in the recent evolution of the doing of UK food and food safety. In the first period (before the mid-1980s), 'food and agricultural systems were regarded as being safe unless scientifically proved otherwise', with farmers and producers powerful and the State having 'a key role to play in the food supply sector' (2009: 124). In the second stage (between the mid-1980s and late 1990s), power shifts significantly as regulation becomes driven 'primarily by the way that food safety issues are perceived by large food retailers; leaving the State to act mainly as auditors rather that enforcers of the mainstream process' (2009: 124–125). Here some of the supply chain management and food standards strategies designed and applied by the large multiple food retailers (of which some background was provided in Chapters 4 and 5) become increasingly significant. Finally, post the BSE-crisis and marked for Marsden et al. (2009) by the establishment of the Food Standards Agency (FSA) in April 2000, a third phase emerges and is storied as the protection of consumers and their interests. Responding to a perception captured in the 1997 James Report, that the existing governance of food safety in the UK was characterised by powerful farming, producer and retail groups, and that decisions were made with a marked lack of transparency, the new Labour government shifted key responsibilities regarding the setting and enforcement of food standards from MAFF (the Ministry for Agriculture, Fisheries, and Food (later Defra)) and the Department of Health to the new and more 'independent' FSA (Marsden et al., 2009: 125).

Marsden *et al.* characterise this current stage of UK food regulation symbolised by the operation of the FSA as:

- a Europeanisation of food policy in the UK;
- the relative empowerment of consumer groups with respect to food safety issues; and
- the further influence of major food retailers in shaping UK food policy.

They refer to the outcome of these three characteristics as the 'hybrid model', and further specify hybridity as being:

> …a response to the (post BSE) food regulation pressures in ways which safeguard the broader macro-economic and political concepts of the European Internal Market (EIM) and increasingly integrated exchange of food goods within and beyond Europe, while also simultaneously enunciating new standardised and 'non-competitive' controls in the name of the European consumer and public interest (Marsden et al., 2009: 143).

We might summarise the shifts Marsden et al. (2009) trace as a story of how food regulation becomes ever more a security problematic in the Foucauldian sense of security that was fleshed out in Chapter 2. First, in the sense that the 'hybrid model' involves both the quantities and qualities of edible matter increasingly being figured and enacted as a matter of circulation; indeed matter is increasingly figured and enacted *as* circulation. The 'increasingly integrated exchange of food goods', which Marsden et al. detail, captures the extent to which food is expected to travel more widely and quickly to service expanding productivity. Or as the opening line of the EU's general Food Law (adopted in 2002 and effectively the umbrella legislation organising all regulation of food safety) puts it:

> The free movement of safe and wholesome food is an essential aspect of the internal market and contributes significantly to the health and well-being of citizens, and to their social and economic interests.

Second, Marsden et al. note that this free circulation of food is accompanied by 'enunciating new standardised [food safety controls] controls'. This reminds us that a key insight of Foucault's work on circulation was his recognition that it does not simply facilitate the movement of 'goods' in either sense of the word. Instead, as the UK's food supply chain has expanded in its range, extended in its length and intensified in its complexity, it has found itself encountering 'new' food safety risks and dangers (including but not limited to BSE, *Salmonella*, *E. coli*, *Campylobacter* events). If the kind of pandemic-triggered 'crisis of circulation' mentioned in Chapter 1 is to be avoided, then food safety regulation has to be adequate to this world of extended and intensified circulation. The FSA's (2011: 20) initial strategic objective gives a good flavour of this when it states: 'Food safety is our concern from the time ingredients come into the country, or leave the farm, right through to when food is sold to you in shops or served to you if you are eating out.'

Third, Marsden et al. (2009) observe that the 'increasingly integrated exchange of goods' and the 'new standardised controls' take place *simultaneously* (2009: 143, emphasis added). This underlines Foucault's key point that the primary task of 'management' for security involves regulating circulation by qualifying both the good and the bad, maximising the former and minimising the latter. In this case maximising good circulations ensures that food can travel in a 'common' market, from farm to fork, while minimizing those circulations that enable the movement of food-borne disease. Much is expected of food – and food regulation – and all of this provisioning and regulation must take place at the same time.

Ensuring, assuring and insuring the safety of the increasing volume and diversity of food circulations has required increased (re)organisation of those circulations. As we noted in Chapter 2, a key feature of such efforts to secure the safety of food circulations has been the centrifugal character of regulation. State responsibility is now supplemented or even displaced by actions from producers, retailers, and consumers.

Central to facilitating this outward-shifting aspect of security work has been the technology of 'traceability' (see Popper, 2007 for a history). Our research on food safety focused on meat-borne disease given that it was the primary concern of the regulatory bodies we were working with, as well as a major vector for food-borne and zoonotic diseases. Traceability is the reason that the fieldwork situations that we explored were full of marks; marks on the meat in the kitchen (health stamps), the marks that travel with the meat (such as identification stamps on the boxes), and marks that are about the circulation of meat (the invoices, receipts, phone numbers). These marks are part of a broader formalisation of food safety, in which food commodities and those that produce, pass on, and receive them must be certified and recorded (European Commission, 2002, 2004a, 2004b). Such accompanying signification is now required to make meat legal, and those tasked with assuring food safety are obliged to verify the presence of the relevant evidentiary materials and act on their absence as appropriate (Food Standards Agency, 2009, 2010b). This system of traceability has been rolled out in the EU over the past couple of decades, amidst ever-growing worries about emerging infectious diseases and how food may be implicated in their gestation and spread. Meat is being made informed in new ways (on informing material, see Barry, 2005; Bensaude-Vincent and Stengers, 1996). As it becomes more 'articulate', it gathers traces of its journey and transformations from farm to fork. The requirements for and technologies of identity preservation vary between animal types, but for ruminants such as cattle and sheep, it generally goes something like this: first ear tags and electronic identification devices are affixed to the animal within a certain period after birth; next the movement of animals between farms, market and slaughterhouse are recorded in farm records, on an official database, and in passports and food chain information that travels with the animals; then, after slaughter, each carcass is given a new official identity and/or health stamp by the meat hygiene inspector, which is recorded in a database; and finally, as the carcass is distributed and processed further in factories and then at retail, these and additional stamps, receipts and invoices are used and recorded to keep track of the meat one step forward and one step back in the food chain.

Through such techniques and practices, the identity of the animal is preserved with the meat to the point of retail. Although increasingly popularised as a technology of assurance in the hands of some consumers and indeed retailers (for example being certified provides confidence to others in the safety of the food being transacted), in a strict legal sense this system of traceability is designed and made obligatory as a tool of recall and reconstruction (Torny, 1998). It is a means to both establish the origin of an outbreak of disease and recall any affected goods:

> If a food safety emergency occurs, the food can be tracked backwards or forwards through the food chain. This information can be used to withdraw or recall food more quickly from the market and to target these actions to specific products (Food Standards Agency, 2010b: 13.2).

In this role, traceability is not really designed to prevent things going wrong, nor to reassure us that things are not likely to go wrong, but is infrastructure working to aid rapid and robust response when something does go wrong, and acts to disincentivise wrong-doing by theoretically, at least, allowing for the identification of malpractice.

Recent developments emphasise that both these centrifugal tendencies characterising food safety governance within the UK (redistribution of responsibilities and the traceability agenda made possible by informed materials) are extending and intensifying their reach. The changes recommended to meat inspection regimes throughout the EU made by the European Food Standards Agency (EFSA) in 2013 (and still under review at the time of writing) exemplify this process. Put simply, these propositions involve the abandoning of routine physical inspections of carcasses in the name of 'modernising' an archaic system of meat inspection dating back to the early twentieth century. Traditional inspection regimes centre on signs of disease or animal welfare issues exhibited before slaughter and on the physical integrity of the carcass. The live animal and carcass are inspected for cleanliness and signs of health and, post-mortem, the carcass is opened as a matter of course, certain organs palpated, incisions made in flesh, tissues checked for damage or lesions, and the head removed and split; all practices designed to identify visual and physical indicators of ill health with negative consequences for humans if consumed. The argument made by those proposing changes is that such an approach is not adequate to the key matters of concern requiring surveillance. In the contemporary food chain these concerns are primarily cellular and molecular and unlike worms or other visible parasites that had traditionally been the foci of attention, these conditions are often impossible to detect on site through visual inspection. Visible and physical characteristics that were useful proxies for the suitability of meat for human food systems in the past are now, it is argued, better assessed through assays of microbiological co-travellers.

On this basis, existing versions of meat inspection are to be replaced by a new approach that is underpinned by two central tenets. The first tenet is increasing and generalising traceability through more thorough food chain information (FCI). The latter will incorporate data on where an animal was born, any movement that may have occurred during its lifetime, and veterinary or other interventions. The second tenet is the use of this enhanced FCI to predict risky individuals or cohorts of carcasses (where risky means more likely to be biologically or chemically contaminated), which would be the target of largely industry-led inspection practice and primarily based on microbiological sampling and testing. Thus, substantial shifts are taking place both in inspection practice and the logics organising meat inspection.

These tendencies (towards food (safety) being considered as a matter of circulation), developments (the redistribution of responsibility for assuring food safety and the technologies of traceability that underpin it), and intensifications (towards a risk-based use of Food Chain Information) offer compelling evidence that a centrifugal-security analytic is indispensable for thinking through UK food safety regulation and how it is patterning disease situations. However, it is our proposition that to proceed as if this particular diagramming of food safety has simply replaced (or could simply replace) what came before would be to misunderstand both the specifics of food safety situations in the UK, and the more general manner in which diagrams inter- and intra-act in their spacing of the social. And that is because – for a very basic legal reason – another diagram, another way of organising the discourse and practice of food safety remains, and must remain, in place in the UK as elsewhere. It is a diagram that brings another set of spatial strategies and operations into the mix of how food is currently kept safe.

Key to this other diagram is the simple fact that food safety remains defined both globally and within the UK as an 'absolute', a 'non-negotiable' quality in the sense that food business operators (FBOs) are prohibited from passing on any foodstuffs to consumers that may be 'injurious to health' or otherwise 'unfit for human consumption' (UK Government, 1990: s8.2). This is how the notion of the safety of food has been enshrined in UK law for 150 years, and what it means is that food safety cannot be managed solely through an approach that relies on risk-based modulating interventions in complex circulations of the sort that, as we have seen, are the basis of the security diagram as described by Foucault. Instead, the notions of 'strict liability' for passing on unsafe food and the 'imminent risk' with which inspectors assess the likelihood of the same mean that the regulation of food safety stands in marked contrast to related arenas like health and safety legislation, where risk-based regulation can take a more thorough-going grip on procedures and practice (Demeritt et al., 2015; Howard, 2004). What this need to protect consumers (and – in reputational and legal terms – themselves) has meant for the large, integrated food multiples in particular is that another set of logics and practices co-exists with the centrifugal ones surveyed above. These logics are guided by the need to keep food safe from external contamination and adulteration (see Chapters 2 and 3).

An example would be HACCP, a process for identifying specific sites of hazard in any given production process, and then putting in place solutions for their control, which are then monitored. As Demeritt et al. (2015: 8) note:

> EU law demands not just clean premises and appropriate food handling and storage, but also the use of the Hazard Analysis and Critical Control Points (HACCP) food management system to address each potential hazard critical to food safety (Directive 93/43/EEC). While some commentators describe HACCP as a risk-based system (Hutter, 2011), its purpose is to eliminate hazards 'in order to guarantee safety' (European Commission, 2009); it does not prioritise control measures based on the likelihood and consequence of exposure to hazards or take any account of the marginal costs or uncertainties of control measures. Indeed, the safety focus of HACCP is best illustrated by its original development by NASA to ensure that astronauts were not poisoned by the ready-meals they took into space.

In other words, there is more than one diagram currently organising the regulation of food safety in the UK. The moves to redistribute responsibility for keeping food safe and the techniques and technologies of traceability described above can fairly unproblematically be characterised as a means of facilitating current and future risk-based interventions designed to modulate the elements of a primarily circulatory social by spacing attention centrifugally (and thus typical of a security diagram in Foucault's sense). However, the sorts of actions that Howard highlights operate in rather different modes. The standards of 'strict liability' and identification of 'imminent risk' with which Environmental Health Officers (EHOs) and Meat Hygiene Service (MHS) inspectors are required to work, and the immediate closures and possible prosecutions which are the consequences for transgressing those standards of hygiene (see below for more details and also (Hyde, 2015)), are examples of straightforward prohibitions and punishments much more characteristic of the sovereign-legal diagram identified by Foucault and which (as was detailed in Chapter 2) operates primarily as far as disease is concerned in terms of spatial division and exclusion (rather than the modulating interventions in circulations of a security diagram). And the HACCP system which has become so key to the conduct of food safety in the UK as elsewhere, despite (as Howard notes) its frequent framing as a risk-based intervention, actually operates through a classic disciplinary diagram in the sense that it is an attempt to institute a series of checks and controls that will prevent a breach in (food safety) regulations before they happen and thus eliminate such hazards from the system. Significantly for our interest here, in contrast to the centrifugal distribution of attention and intervention demanded by the figuring of circulation as the space of action in a security diagram, both the diagrams of sovereignty and discipline involve more centripetal, inward orientations of attention, whether it be the case of the bounded space of the (clean) territory (in the case of hygiene standards), or the controlled space of the food chain (in the case of a HACCP mapped food business).

As we noted in Chapter 2, the fact that new diagrams do not simply supersede and replace the old was well recognised by Foucault (2007: 8), who prompts us to recognise that in any given situation we are likely to encounter the operation of more than one diagram, and to specify the configuration by which they more or less successfully co-exist. What might we say then about the current specific configuration of food safety regulation in the UK?

First, that whilst recognising the presence of the other diagrams at work in the shaping of food situations is critical in a number of senses, it is undeniable that in the current configuration it is security which is in the dominant position, and further that in this configuration characteristic features of the other diagrams become captured and translated by that hierarchical relationship (Bingham and Hinchliffe, 2008; Lemke, 2015). In particular, the controlled spaces shaped by the disciplinary diagram are transformed from isolated environments to nodes in a wider system of circulation, and the borders of the sovereign-legal diagram are transformed from hard and fast perimeters into something more like monitoring mechanisms. In more practical terms, food businesses shift from being (enacted in a disciplinary mode as) 'floors, wall, and ceiling' (to use 'inspector speak' for older regimes of practice) to being (within a security-led configuration of diagrams) 'sites' within a broader set of flows. They shift from being a bounded space (in a legal-sovereign mode) where everything of interest is present, to being (within a security-led configuration of diagrams) a space where marks and other trace-work are used to confirm the site's successful intrication with the outside.

What we might further propose about the current specific configuration of food safety regulation in the UK on the basis of its configuration as a hierarchical arrangement of security, disciplinary and legal-sovereign diagrams, is that it is given important figurative shape by the image of the 'food chain'. Or rather a very particular version of the food chain (Loeber, 2011). Clearly in its use to represent the 'journey' of food from 'farm to fork', the food chain will always highlight first and foremost the circulatory aspects of food production. But we would suggest that what is distinctive about the version of food chain that currently helps to organise the UK food regulatory system is that it represents the chain of links more as a system of interconnected tubes through which food is transformed and transported between a series of key nodes in the food production network, tubes that act to isolate this work and movement from the dangers of the outside world (coded dangerous and dirty) and are oriented to a goal of food being safe at the point of sale that is demanded by current food law. In this sense, it mirrors directly that image-dream of the vertically-integrated food production system detailed in Chapter 4. It conceals rather than foregrounds the work of coordination, smoothing out a series of diagrams and associated spatial tendencies and operations. As such it leaves us with sites, as linked nodes in a conventional network, rather than situations as topologically complex meeting places at which not everything is necessarily straightforwardly present. This smoothing out has – as we shall see – some important consequences.

A Failure of Coordination?

Just as the EFSA-proposed changes to the meat inspection regimes were being reviewed and the process of translating them into country-specific practice was beginning, an event occurred in 2013 that served to very publicly problematise the current configuration of food safety regulation in the UK, including the future direction flagged in those changes. Though certainly not a 'crisis of circulation' of the kind figured in Chapter 1, nor even strictly speaking a food safety issue at all, the aforementioned horsemeat scandal represented enough of a breakdown of the operation of prevailing UK food safety governance for the workings, failures and limits of that system to be exposed and held up to critical scrutiny.

Briefly put, in early 2013, processed meat products sold as beef-based in several UK and Ireland supermarkets were found to be adulterated with as much as 100% un- or improperly declared, criminally-introduced horsemeat. This situation was subsequently found to be widespread and resulted in a series of investigations and reviews including the Troop Report, the National Audit Office Report on food safety and authenticity, the Environment, Food and Rural Affairs Committee (EFRA) Report into Food Contamination, and the Elliott Report led by Chris Elliott, a Professor of Food Safety. Although there was little suggestion that human health had been compromised via this adulteration, the event served to draw attention to the serious limits of food safety regulation defined by the new EFSA agenda and its key tenets.

First, traceability was a key focus of the reviews. There were repeated concerns expressed that there had been an apparent failure of the system of labelling and that businesses and regulators had not been able to detect that beef had been adulterated with horsemeat. Labelling requirements for beef had been introduced following the disruption in the beef market caused by BSE before later regulation extended such traceability to all food businesses. In practice, these regulations compel those in the food sector to be able to identify the supplier of any ingredients they work with, and the destination to which the material produced is sent. This is known as the 'one step back, one step forward' approach. That this apparently rigorous process failed to identify such a widespread issue of adulteration raised questions for many about the effectiveness of the system. Felicity Lawrence (2013) of *The Guardian* was more blunt:

> The industry has previously boasted that it has full traceability of its supply chain which it audits frequently. The current scandal shows that that traceability is not worth the paper it is generally written on. Most of the factories caught up in the scandal have accreditation with mainstream auditing schemes such as that run by the British Retail Consortium but it failed to spot the problem.

More precisely, longstanding concerns about how traceability operated for processed meats appeared to have good basis. The horsemeat which found its way into the food chain was from 'filler' products added to economy ranges. The product is manufactured by the separation of lean meat from fat and connective tissues and was delivered to the burger manufacturers in frozen blocks. The filler does not count as a meat ingredient, although economy burgers are only required to contain 47% beef. As such, there could be no way of identifying through physical examination the species of the meat product that was provided; the label would be the only evidence apart from a species test.

Second, the limits of the risk-based microbiological testing were demonstrated. What became apparent in the fallout from the horsemeat event was that however extensive a regime of food sampling had been, the adulteration of beef products with horsemeat would not have been spotted in the UK. This is because a test for horsemeat was not undertaken in regular testing practice and there were no plans to make an exception to this. As Catherine Brown, Chief Executive of the Food Standards Agency (FSA) put it:

> We have accredited tests, and we have a mixture of DNA and other tests that we could use. We have tests available so that, had we tested and had there been the wrong stuff there, we would have found it. The real issue is that we would not have tested, because our surveillance approach is risk-based and largely intelligence-led. This year, we are focusing on formaldehyde in kitchenware, contamination of baby milk, false vodka that makes you go blind or kills you, and *E. coli* in sprouted seeds, amongst others (Environment Food and Rural Affairs Committee of the House of Commons, 2013: Vol II Ev.5).

Similarly, the internal food testing regimes of retailers are based around past experience. As such, testing practice almost inevitably follows situations already experienced, meaning that novel acts of adulteration will seemingly evade detection. The limits that such an approach presents in terms of detecting novel pathogens is clear enough.

Finally, the limits of distributed responsibility and self-regulation within food safety governance were made visible in the aftermath of the horsemeat event (see also Jackson, 2015). Internal sampling regimes were accused of being of questionable efficacy and relevance, with no mandated or even recommended standard in place, questionable targeting, and an overall lack of coordination between FBOs:

> All retailers had their own approach to supply chain control. How they specifically applied that to their own supply chain was a matter of individual choice. The British Retail Consortium (BRC) had no recommended approach for their members. There is evidence that some testing by major retailers may not have been targeted, but provided comfort. For example, the testing of prime beef cuts from single species cattle abattoirs, sourcing cattle direct from farms, for horse DNA was of little value. This lack of coordination was surprising, and a set approach could be helpful. (Scottish Government, 2013: Doc. 7, para 17.3).

The auditing systems designed to ensure the effectiveness of these sampling regimes and other checks on safety also came in for criticism, with Elliott describing the quality of private audits as 'variable', some of their requirements as 'futile', concluding that they are 'not achieving the intended purpose'. More than that he suggests that they may even be counterproductive in that they place 'unnecessary' costs on food businesses, particularly small and medium-sized enterprises (DEFRA/FSA, 2014: 39).

Elliott's criticism echoes academic work on the self-regulation approach which has problematised the assumption that businesses necessarily have the requisite resources to properly understand and manage their risks, a situation which can too often leave those with the finances buying in services that are not appropriate and those without the finances wholly reliant on state regulators to help them (state regulators whose availability, as we will discover later in this chapter, is increasingly limited) (Hutter, 2012: 18).

The point of repeating and rehearsing here the criticisms of the traceability, risk-based sampling, and distributed self-regulation that emerged in the wake of the horsemeat scandal is that together they point to a concern that the direction in which food safety regulation is moving leaves it less rather than more able to deal effectively with precisely the kind of entangled situations in which we have proposed that disease emerges and is amplified. The food adulteration event reviewed here is an example of one such situation, where monitoring regimes made it easy for cost-cutting and criminal acts to introduce illegal (but not in this case unsafe) meat into the food chain without detection. The case of *Campylobacter* within UK poultry production reviewed in Chapter 4 is another, where the pressures of just-in-time regimes generated the conditions of possibility for disease proliferation. In both cases the idea that an integrated and prophylactically secured food chain alone could keep things safe was demonstrated to be not only false but perhaps producing its own biosecurity risks.

What has become clear is the extent to which current UK food safety regulation might be dominated by an approach where organisations are called to account for the actions that were or were not taken to reduce disease risk (or an 'anticipation of retrospection' (Miyazaki, 2003: 259 and Chapter 2)), where responsibility is performed through auditable acts of disease prevention. Elliott is scathing about the worst excesses of this in his review of the horse-meat scandal:

> Risk management is important and starts with knowing who you are doing business with. *Understanding the complexities of supply chains is much more than maintaining a paper trail.* When things go wrong, this alone will not provide a defence against allegations of negligence (DEFRA/FSA, 2014: 20, emphasis added).

It is in this context that Elliott tempers his overall assessment that the UK currently has 'one of the safest food supply systems in the world' (DEFRA/FSA, 2014: 11), with a concern that the resources that support the regulation of food safety (especially at the Local Authority level) are currently being 'cut to the bone' (2014: 14). It is to those resources, their technologies and current fragility, that we now turn. Our key proposition is that inspection – at least in principle – is capable of offering precisely the 'much more' that Elliott suggests is necessary to understand the complexities of supply chains, and that in this context its current marginalisation in favour of a larger reliance on Food Chain Information should be considered problematic in terms of keeping food safe.

Inspection as Tending the Tensions of Food Safety

The kind of work through which we are proposing that food inspectors routinely engage the complexity of food safety situations is not something that can be found in the training manuals and related materials which codify the job of a food or environmental officer. Rather, it is something that we witnessed as part of the fieldwork that informs this book. The particular interest of this strand of ethnographic investigation was the ways in which regulatory requirements for the production and handling of meat were managed at different stages of the food chain to prevent the spread of food-borne diseases. We studied the ways in which farmers, food business operators and food safety enforcement officers negotiate food safety regulation in practice, and (as will become significant in what follows) the interplay with other commitments and concerns in producing food.

Our approach was to follow meat through the various sites as it travels from farm to fork. We shadowed representatives of the three main groups of inspectors responsible for ensuring, insuring and assuring food safety and food standards. These are the Trading Standards Officers (TSOs) responsible for food standards (of whom we had one key informant whom we accompanied on multiple visits), Meat Hygiene Inspectors (MHIs) responsible for conditions at abattoirs and most meat processing businesses (of whom we had two key informants, both Lead Official Veterinary Surgeons (OVSs) whom we accompanied on multiple visits), and Environmental Health Officers (EHOs) responsible for food safety at public-facing establishments as well as some processing businesses (of which we had three key informants whom we accompanied on multiple visits). Over a period of two years we spent time at all stages of the food chain: from farms, agricultural shows and livestock markets (looking at live animals); through to the slaughter houses, cutting and processing plants, and cold stores where the animals are killed and transformed into meat products; at the Border Inspection Posts (BIPs) where food from outside the country enters the UK food chain; and then to the supermarkets, butchers, produce markets, restaurants/hotels/catering outlets and other retail sites that collectively make up the consumer-facing end of the food chain. In addition, we also spent time with DEFRA officials, FSA regional staff, attended FSA training events, and visited one of the Health Protection Agency's food testing laboratories where samples collected by EHOs and MHIs are taken for analysis.

Despite recent institutional rearrangements (for example the Meat Hygiene Service (MHS) becoming part of the FSA in 2010) and increasing national variations within the UK (in the wake of devolution), the aforementioned three sets of inspectors (TSOs, MHIs and EHOs) have remained responsible for the enforcement of food safety (EHOs and MHOs) and food standards (TSOs) for over a century. They are tasked with monitoring 'compliance' with standards required by UK and now EU legislation. However, we found that this narrow description of their work does not do justice to its significance. Instead, our key conclusion from spending time alongside the inspectors was that – at its best – food safety inspection has much to teach us about what it takes to understand and deal effectively with the challenge of the intra-actional and topological complexity of situations, and specifically disease situations. The rest of this section distils our observations into three key insights in an inevitably composite account that seeks to be faithful to much that was common between inspections without losing sight of the important variations between them. The first is that inspectors' *involvement* in food safety situations offers them the basis for understanding and dealing effectively with the complexity of those situations. The second is that inspectors' *appreciation* of the range of things that are demanded of food at a site *and* the vital coordination work that Food Business Operators do to hold those demands together, deepens their understanding of the complexity of food safety situations. And the third is that inspectors' ability to improve food safety situations often depends on their ability to successfully *articulate* the different diagrams that pattern a given disease situation.

We turn first to the importance of inspectors' *involvement* in food safety situations for their understanding of and ability to deal well with them. Such involvement would originate in one of two ways for the inspectors that we shadowed. For some of the officers it began with their mandated presence in certain situations, specifically the entry of animal-related food material into the UK through Border Inspection Posts (BIPs) (which must always be overseen and approved by a member of MHS staff), and the killing of animals for meat at abattoirs (which similarly must always be overseen and approved by a member of MHS staff). More often, however, involvement in the complexity of food safety situations for TSOs, MHIs and EHOs comes through the key technique of the visit.

Although food business operators can be (and increasingly are) contacted by those responsible for food safety regulation in a number of modes – by letter, email or phone – it is the regular and often unannounced attendance 'in person' of an inspector or inspectors at an FBO's place of work on which the food safety system currently depends. Having powers of entry that exceed those of the police in some circumstances and powers of enforcement that include immediate closure of premises means that – even in the hybrid model of food regulation where (large) producers shape the landscape in so many ways – this remains a relationship in which EHOs and MHS personnel can (and do) exercise significant authority. The frequency with which a particular FBO is visited largely depends on the type of food handled and the sort of operations on that food that take place, whilst the precise timing of a visit is the result of a complex mix of factors including risk assessments, current strategic priorities, the schedules of work of particular inspectors, as well as any relevant concerns raised by members of the public that are considered serious enough to follow up.

In terms of what goes on in a visit, inspectors were trained and directed by their handbooks and other paperwork to pay attention to four interlinked elements of the food business: first the animals/meat 'itself', second, the infrastructure of where they were visiting (the aforementioned 'floors, walls and ceiling'), third, the records and marks involved (on and about meat, but also staff qualifications and business certifications), and finally the people and their practices (not just the staff and managers but sometimes also customers or consumers present). This distribution of attention was true whether the inspector was a TSO, MHI or an EHO, although the relative importance given to each could vary depending on the specific circumstance. For example, in a cold store, access to the meat itself tended to be limited by the packaging (and thus paperwork had to be taken as a surrogate for its condition), whilst in an abattoir inspectors are obliged to witness the carcasses in a physical sense.

What was also consistent between types of inspector and inspection was that this four-fold distribution of attention only begins to do justice to the visits. Indeed, it was the work between and through these categories that came to interest us most. Michel Serres's (2008: 304–305) description of a 'visit' as something that involves going and seeing, something that involves the whole body, something that always explores 'a knot', offers a good sense of how we learnt to experience inspections through shadowing our informants. For them the food businesses that we inspected, whether a livestock market, a processing plant or a hotel kitchen, were very much complex presents. While the four strands of the food safety landscape of the business were critical, it was actually the emergent knot generated by these strands that the inspectors attended – and as we shall see – tended.

Such an immersion in the complexity of a food business is not simply or only a choice on the part of dedicated inspectors, although we witnessed an enormous amount of dedication in our fieldwork; it is also in part a matter of necessity. We noted as we shadowed inspectors examining livestock on farms and at markets, handling fresh meat at the slaughterhouse, inspecting restaurant kitchens, and taking samples from farm animals or food products at retail for testing in the lab, that they rarely, if ever, encountered disease directly or immediately. The work of food safety inspection is most often that of looking for subtle signs: the odd behaviours or forms of animals; the unusual appearance and feel of animal flesh and intestines; the presence of viral antibodies in blood samples; the hygiene conditions of farms, vehicles, factories and restaurants; bacterial growth cultured in plastic dishes in the laboratory. In some cases these signs may take the form of what are indeed technically known as 'disease signs' (things that accompany disease, such as weight loss or a lack of coordination of an animal), in others they operate as surrogates (things that stand in for disease risk, such as mouse droppings in a restaurant kitchen).

If inspectors are going to properly assess the state of the food business they are visiting and the likelihood of it being a locus of disease, they have to involve themselves in the situation of the business, or the configuration of numerous and heterogeneous agencies, the complex present of the mingled place of the border inspection post/farm/market/slaughter house/processing plant/cold store/kitchen and the mingled object of the meat – to get a sense of what is going on. They were involved in scanning paperwork, squeezing mouse droppings, scrabbling about on their hands and knees with a torch. Of course, there were important variations here – getting involved in the complexity of a small kitchen that you visit frequently as an EHO because it has been served with an improvement order is very different from getting involved in the complexity of an enormous cutting plant which you might visit once a year at best as an MHI – but our experience was that good inspectors found ways of using diverse sources of information to generate a composite portrait of the many different kinds of FBOs for which they were responsible. Such a picture – it is worth underlining – would typically offer a very different impression of the organisation than that provided by one (or even many) of the audits that are so regularly performed as noted above – on most medium to large food businesses. The multi-sensory involvement of the inspection visit tends to offer a version of the complex present of a *situation*, while the pre-codified formality of the latter can reduce matters to the simplified working of a *site* (to use the distinction from Chapters 1 and 3).

Involvement in the situations of food safety that inspectors gain through their presence at or visit to food business operators might helpfully be thought of in terms of their 'learning to be affected' by their situations (Despret, 2004; Latour, 2004). As such it forms the basis for their *appreciation* of the sometimes very different things that are demanded of food in a given production situation and the vital coordination work that food business operators do to hold those demands together.

What good inspectors learn to appreciate through their involvement is that meat is actively made in practices that start well before and continue well after the death of the animal, and further that these practices are by no means all or even often directly concerned with food safety. Certainly the situations of food production are significantly shaped by the exclusions (such as policed boundaries) required for good hygiene (legal-sovereign diagram), the controls (such as HACCP mapping) required for the safe processing of food (disciplinary diagram), and the monitoring mechanisms (such as traceable meat) required for the optimal circulation of food (security diagram), but attendance at or a visit to a food business quickly makes clear that other diagrams working through other practices to achieve other qualities of meat than safety are also at work. Stocking practices, for example, are organised through an economic diagram which seeks to make meat production as profitable as possible for the large supermarket chains (Chapters 4 and 5) (Marsden et al., 2009). Practices of judgement associated with religious-cultural diagrams are mobilised in connection with determining appropriate stun levels for halal-compliant meats (Miele, 2013). Culinary-taste practices associated other more secular cultural diagrams such as the hanging of duck carcasses in the kitchen before their preparation in certain restaurants and other consumer-facing establishments. These practices all enact, or realise, different versions of meat at different points in the food system; as profitable, as culturally appropriate, as tasty, depending on the diagram involved.

More than this, the inspectors that we shadowed had been forced to appreciate that the practices, versions and expectations associated with different diagrams are not always or even usually separated and distributed through the food chain. Instead more than one is most often in play at any time and place. In terms of our fieldwork focus, meat (certainly after slaughter) is always expected to be safe wherever and whenever it is. If it is being assessed by a supermarket buyer, it is also expected to look good (enough). If it is being bought in a halal butcher, it is also expected to have been treated well (enough). If it is being served in a restaurant, it is also expected to be tasty (enough).

These practices, the logics they enact and the versions of meat that they produce can relate to each other in very different ways. They may simply co-exist with little or no obvious mutual impact. They may resonate as in the case when we heard a Meat Hygiene Service inspector enthuse when told by a Food Business Operator that they were using traceability techniques required under food safety legislation to improve their stock control. Or they may interfere and pull in different directions, as in the case when we witnessed an Environmental Health Officer declare that leaving the aforementioned duck carcass hanging in the middle of the food preparation area in the interests of better taste was a food safety hazard and must be halted at once. Or when we witnessed frustration (although grudging compliance) on the part of an FBO with how the requirements of in-situ inspection and testing of carcasses at abattoirs slows down the circulation of meat for processing at a critical point (and so affects the profitability of the plant).

In these latter two cases different diagrams shaping the same situation were generating tensions.

It is at this juncture that the question – and the challenge – of management come more clearly into view for food safety inspectors. For if the simultaneous presence of different diagrams, expectations and practices mean that situations of food production, the activities of food business operators, and indeed the food itself, are all literally matters of coordination, then that coordination and how it is achieved matters very much in terms of keeping food safe. Precisely because the situations, activities and meat are not – and cannot be – all about food safety, recognising and supporting the work of coordination (Mol, 2002) that FBOs do to get these multiplicities to hang together – and handling the tensions that this involves – becomes a critical intervention on the part of food safety inspectors. For it is precisely when divergent diagrams, logics and practice are not acknowledged as such, precisely when the resulting tensions are not addressed, that the intensities that generate the conditions of possibility for disease outbreaks occur (for other similar examples and arguments see Hinchliffe, 2001; Law and Mol, 2008).

Once again, there are significant variations between how and the extent to which different inspectors in different food safety situations are able to appreciate the diagrams that pattern the food and the resulting coordination work this might demand of FBOs. The full complexity of a business might be relatively easily accessible to a good MHI if they are a regular visitor to and have a good relationship with the management of a small, sausage-making firm. Gaining the same direct insight into what else other than food safety requirements is shaping the meat through a visit to a large multiple-site supplier of processed products to a large supermarket chain is a very different proposition. But, once again, we were continually struck by the ability of inspectors to piece together or construct a patchwork (Chapter 5) of issues in order to understand the food situations to which they attended. For example, at an otherwise unremarkable visit to a cold store, we witnessed an MHI able to offer culturally-specific advice to the manager about how best to impress the delegation of Chinese officials who were soon to assess the site for approval for holding meat for certain of their companies.

Just as the involvement of food inspectors in food safety situations formed the basis for their appreciation of both the multiple diagrams that pattern food and the coordination work this requires from FBOs, so in turn that appreciation is the basis for inspectors' ability to improve complex food safety situations through *articulation* of the logics and practices of those diagrams. The support that we witnessed inspectors offer all sorts of FBOs in the service of food safety was a striking and – we feel – important feature of their work and our fieldwork (for more on tending the tensions of food safety, see Bingham and Lavau, 2012).

Most obviously, the inspectors that we accompanied were very aware of how the food businesses that they inspected were part of a broader economy of exchange, qualities and transaction. For an EHO visiting a restaurant, for example, this broader economy would include the local council itself, through the rates that businesses pay, and through its economic strategy to support the performance of the local economy. We heard again and again from EHOs, TSOs and MHIs that they were not looking to close down or unnecessarily penalise the businesses that they attended. Certainly, if they encountered what is technically known as 'imminent risk', then they would not have hesitated to enforce the law promptly and strictly. And if food-borne illness was associated with a particular business, then their priority would certainly have been investigation and control (see also Hyde, 2015). But most of the time – and certainly in the majority of the visits where we accompanied inspectors – the focus was on tempering enforcement with education and enablement, whether that was done by supplying guidelines and advice to those starting out, or producing action lists from their inspections to managers with timetables for remedies rather than forcing businesses into inactivity. The concern thus was with on keeping food safe by improving performance.

An example: in a restaurant visit accompanying an EHO, various food handling and documentary offences (e.g. a wooden chopping board, mouse droppings behind workstations, and lack of comprehensive HACCP records) are identified and yet the chopping board was not immediately confiscated (rather the manager was asked to switch to an approved type), because the mouse droppings were not fresh and were 'only' on the floor and not on the food storage shelves the business was not immediately closed on the grounds of imminent risk (rather advice for how to – quickly – restrict access to rodents was offered), and food preparation was allowed to proceed even in the absence of evidence of a full HACCP mapping (rather another information pack about food safety management systems was immediately promised to help with the paperwork). The important work here is done between the EHO and the manager. What we witness is her making use of the 'discretion' (Hutter, 1988; 2011) that is so critical to this kind of work. Not discretion about whether to enforce, but how. As Hutter summarises:

> EHOs typically hold high levels of discretion about how to implement the law in individual premises and the evidence is that they deploy this discretion flexibly. Their overall approach does not take enforcement of the law to refer simply to legal action; rather it refers to a wide array of informal enforcement techniques such as education, advice, persuasion and negotiation. Securing compliance is its main objective, both through the remedy of existing problems and, above all, the prevention of others. The preferred methods to achieve these ends are co-operative and conciliatory. So where compliance is less than complete, and there is good reason for it being incomplete, persuasion, negotiation and education are the primary compliance methods. Accordingly, compliance is not necessarily regarded as being immediately achievable; rather it may be seen as a long-term aim. The use of formal legal methods, especially prosecution, is regarded as a last resort, something to be avoided unless all else fails to secure compliance. Indeed, the importance of legal methods lies in the mystique surrounding their threatened or possible use rather than their actual use (Hutter, 2011: 71).

Our inspector, then, has some wiggle room herself and offers it to others as she sees appropriate. She uses materials as boundary objects 'both plastic enough to adapt to local needs and constraints of the several parties employing them, yet robust enough to maintain a common identity across sites' (Star and Griesemer, 1989: 393). This is the borderlands of inspection and involves deploying a range of modes and manners to achieve the aim of compliance and improvement both within and between visits. She modulates her manner with the manager: while she is being stern with him, she is also being supportive, advising him to contact the pest control company, explaining why wooden surfaces are less hygienic than stainless steel. Working skilfully, she forces the manager to confront and attend to the different versions of meat entangled in the situation of the kitchen: sorting between those versions (asking for explanations when the reasons for doing something a certain way in the given situation were not clear); testing their compatibility (for example, establishing whether the level of training of the staff were consistent with the minimum legal requirements); identifying their contradictions (for example, pointing out that using cheap suppliers may save them money but may be exposing them to risk on the food safety side); ranking their priority (being very clear that if an inspector found mouse droppings on food preparation or storage surfaces they would be closed, lunchtime or not); working with their resonances (explaining that keeping good records will help them be more economically efficient and meet their food safety responsibilities).

This is inspection as articulation work. Articulating first as linking together the different diagrams that constitute the current conjuncture of food safety, helping food business operators manage the legal-sovereign work of hygiene, the disciplinary work of hazard control, and the security work of circulation tracing that is required of them, with all the different objects and directions of attention this necessitates. Articulating second as weaving tightly enough the different qualities as well as safeties that FBOs require food to be, helping managers maximise the positive resonances and minimise the negative interferences between the imperatives dictated by all the different diagrams shaping the situations of production such that their need for their food to be tasty, good value, ethically sourced, and so on do not pull in opposite directions to the need for safety. And articulating third as generating the knowledge/practices by which the safety of food businesses might be improved by storying them in the presence of and ideally with the participation of those invested in those businesses, helping managers by using discretion and diplomacy in the first instance that shows an appreciation of the coordination work they are always doing.

Key to all these articulations by food safety inspectors is a taking seriously of the many different things that food is expected to be, the many different forces that shape the situations of food production, and the many different practices that are involved. What they offer, taken together, is not the official 'smoothed out', tube-like version of the food chain reviewed earlier in this chapter but instead a much more topologically complex one that does justice to the knots and complexities of the supply chain (to return to Chris Elliott's point earlier). And as such they offer an opportunity for 'tending the tensions' of food safety, the etymological and practical link between extension, intensity, tensions, and tending is made by philosopher Jean-Luc Nancy (2008: 134), such that 'consequently there are possibilities for ethical developments that we might perhaps not expect to find here' (ibid). If this is the case, then perhaps we might think of the at/tending craft/work of food safety inspection as a sort of 'health check' on the complex present of (what we inadequately refer to as) the food chain; an orientation to and perhaps early warning of when pressures are becoming intensities and when intensities are reaching a tipping point.

The suggestion here is not that all EHOs, MSIs and TSOs consistently improve the food safety situations they encounter during inspections through the kind of articulation work described above; we witnessed its efficacy as a strategy mainly in the small to medium-sized businesses we attended during our shadowing rather than in the large multiples. It is also not that all food safety inspections are currently done well. And it is certainly not that inspection by involved, appreciative, articulate TSOs, MHIs or EHOs is all that is needed to ensure food safety. Rather the proposition is that food safety inspection at its best offers lessons for attending, tending and improving the complex situations that constitute the food chain that other technologies do not. If that is so, then its current marginalisation should be a matter of profound concern.

Being Stretched

This marginalisation has been building for some time. In institutional terms none of the key bodies involved are thought to be particularly satisfied with the current distribution of responsibilities between themselves and the local authority inspection teams. As Hutter (2012: 14–15) has recently summarised, from the EU perspective:

> ...the institutional division of responsibilities for policy and enforcement of food safety and standards between central and local government is unusual as, with the exception of the Meat Hygiene Service, the FSA has no enforcement arm...

from the FSA perspective:

> This arms length arrangement presents the FSA with many challenges when directing the activities of individuals they neither employ nor fund, [and...] EHOs themselves have been criticised for being a source of inconsistency, particularly regarding their levels of expertise – as aspects of their remit have become increasingly technical, the adage that they are 'Jack of all trades and master of none' has become ever apparent.

whilst from the EHO perspective:

> …many local authority officers viewing [the FSA] as overly bureaucratic, impractical and difficult.

Recent developments have tended to put the quality of EHOs and MHS inspectors, their expertise, and their inspections under further pressure. As noted earlier, EFSA proposals are for a shift from universal MHS inspection of carcasses to a more risk sample-based approach. Concerns with the de-skilling that is thought likely to be the result have been widespread. Whereas currently trained veterinarians have a presence on slaughter lines, the new proposals allow less qualified members of staff to undertake inspections (see for example, Unison, 2013). As for the EHOs that we shadowed, who, it should be noted were all food safety specialists with postgraduate microbiology qualifications, a term that came up repeatedly in our discussions was 'stretched'.

This stretching spoke to a number of issues. First, stretched between different objects of attention. In the sense that a particular set of regulatory requirements has meant that in inspections on farms, in slaughterhouses, in restaurants, and at ports, more and more attention (and a higher and higher proportion of attention) is being spent on the records and marks that underpin traceability schemes. There are questions being raised by some very experienced inspectors about whether these regulatory requirements for more and more monitoring of records are indeed generating or amplifying bio-insecurities themselves. This is not to say that inspectors were arguing that auditing records does not have any value. Rather their concern was with what happens when auditing records happens at the expense of inspecting animals or food, etc., and when good record-keeping is taken as a surrogate for good animal husbandry or safe food handling. This is partly a matter of time: as the list of record checks lengthens, there is less time for the inspector to attend to other sources of hazard that are not so formalised, such as signs of vermin in animal feed, or evidence of cross contamination in a kitchen. It is also a proposition that a records-based approach may be insufficient in principle for identifying certain food safety issues, whether these are 'classic' matters of concern such as parasites or 'new' microbiological ones.

Second, inspectors are feeling stretched between their priorities. We have already noted how EHOs at least are working at once for the FSA (with its mandate to regulate food safety in the service of public health) and their local authority (who benefit from having open and not closed businesses). More broadly, several of the individuals that we shadowed talked of having to manage the pull between the often competing 'demands' of the Pennington Report (Pennington, 2009), which emphasises the value of 'on the ground' specialist food safety officers, and the Hampton Review (Hampton, 2005), which advocates reducing the burden of regulation on small to medium-sized businesses.

Third, inspectors are increasingly stretched in terms of time, space and resources due to changes in regulatory emphasis and reductions in staffing levels associated with austerity measures in the UK. Whilst consistent and comparable data on total numbers of inspectors and inspections are hard to come by, figures collected and analysed by the criminologist Steve Tombs suggest that in the period between 2003/2004 and 2012/2013 food hygiene inspections in the UK declined by 12%, food standards inspections by 34%, and prosecutions associated with either of these by 28%. Tombs presents these findings alongside data for both Health and Safety Executive and Environment Agency regulation and enforcement over the same period and notes that:

> What *is* remarkable, certainly for a set of social scientific data, is that *each set of data reveals precisely the same trends*: that is, notwithstanding variations across regulators, the form of law being enforced, and indeed within regulators and specific forms of enforcement activity by year, *each set of data unequivocally indicates a long-term downwards trend in every form of enforcement activity* (Tombs, 2015: 154, original emphasis).

Tombs proposes that such a trend, which – as he notes – is least dramatic in the case of food safety, possibly as a result of its politically highly sensitive character, is in large part the ongoing consequence of a politically-motivated, business-friendly 'Better Regulation' agenda, which he translates as 'regulation without enforcement'. It is interesting to note that the dedicated food safety team in the city where our fieldwork was focused has since been disbanded due to these cuts in funding, with a massive loss of expertise. Those that remain are expected to make more visits spread over larger areas in shorter periods. As more than one of the inspectors that we spent time with commented, this is hardly conducive to the appreciation of and involvement in food businesses that they associate with tending the tensions of food safety.

If the success of this work by EHOs, TSOs and MHS officers is dependent on their having the time, expertise and will to appreciate, get involved in and seek to better articulate the complex present of the situations of meat production, what will be the consequences for food safety of such a series of moves within and around the institutional arrangement of food safety which risk profound and irreversible erosion of the skilled practice on which that work depends? These moves include the proposal (and it is nothing more than that at the time of writing) that the principle of universal physical inspection of food businesses should be abandoned altogether in favour of a more targeted or risk-based approach. The answer is as yet unclear, but what is clear is that the relations linking food businesses and food safety regulators are highly vulnerable in all kinds of ways.

Conclusions

This chapter has explored the question of how disease situations are being patterned by the logics and practices of food safety regulation. The quick answer would be 'in a number of ways' to reflect the variety of forces and orientations that were found to be shaping how food is currently regulated in what we were reminded by Chris Elliott remains one of the 'safest food supply systems in the world'. But that same 'in a number of ways' also serves as the starting point for a slightly longer summary of how the preceding material has progressed the argument of the book. For, to be more specific, what we have described is how at least three diagrams currently configure what is expected of food business operators in terms of keeping their product and places of work safe; a sovereign-legal diagram that continues to work by means of hygiene and exclusion, a disciplinary diagram that works by means of mapping and control, and an increasingly dominant security diagram that works by tracing circulations and distributing outwards responsibility for keeping those circulations safe. The way that we have suggested that the current settlement of these 'number of ways' operates is by fudging their significant differences through a versioning of the food chain that effaces two critical matters. First, the topological complexity generated by the play between more regional, more networked, and more circulatory spaces. And then second the vital coordination work that must continually be done by Food Business Operators if food is going to hold together as a successful multiple that manages to meet the various expectations that are made of it.

The (rather desperate) hope seems to be that further boosting the dominance of the circulatory diagram through technologies such as food chain information will solve all issues and avoid the need to address such matters of composition. However, given that we have also learned that there is good reason to suspect that it is precisely the fudging and effacing of complexity and coordination that risks generating the intensities which characterise many disease situations, it seems unlikely that such a strategy will succeed in the long term. In that context it seems unfortunate to say the least that a key resource for helping FBOs manage the challenges of keeping food safe – namely the assistance of EHOs and MHS staff – is being withdrawn in the name of progress and a re-regulatory (Tombs, 2015) agenda. As Hutter (2012: 19) comments:

> ...in the case of food governance as we are also witnessing a situation where the technical demands contemporary food laws place on business are increasing and the traditional knowledge providers for large numbers of small and micro businesses in the UK are EHOs whose numbers are being reduced. Many of the alternative sources of support are costly and beyond the reach of these businesses.

We might add that given the apparently profound limitations in audit culture's ability to understand the complexity of food safety situations identified by Elliott and others following the horsemeat event, it might not only be small and micro businesses that could benefit from the craft knowledge of inspectors.

The point that we would take forward from this into the rest of the book is not simply that the reductions in expertise by specialist food safety personnel in food safety should be reversed (although there may well be a case for that), but that the expertise in articulation they can embody, and the taking seriously of the topological and coordinative complexity of food safety situations that they demonstrate, may offer broader lessons in terms of keeping life safe. For – just like disease – safety is a relational issue and as such the matters of composition it addresses require tending so that they do not fall apart.

References

Barry, A. 2005. Pharmaceutical matters: The invention of informed materials. *Theory, Culture and Society*, 22, 51–69.

Bensaude-Vincent, B. & Stengers, I. 1996. *A History of Chemistry*. Cambridge, MA: Harvard University Press.

Bingham, N. & Hinchliffe, S. 2008. Mapping the multiplicities of biosecurity. *In*: Lakoff, A. & Collier, S.J. (eds). *Biosecurity Interventions: Global Health and Security in Question*. New York: Columbia University Press/Social Science Research Council.

Bingham, N. & Lavau, S. 2012. The object of regulation: Tending the tensions of food safety. *Environment and Planning A*, 44, 1589–1606.

DEFRA/FSA 2014. Elliott Review into the integrity and assurance of food supply networks: final report https://http://www.gov.uk/government/publications/elliott-review-into-the-integrity-and-assurance-of-food-supply-networks-final-report Accessed 30 October 2015.

Demeritt, D., Rothstein, H., Beaussier, A.-L. & Howard, M. 2015. Mobilizing risk: Explaining policy transfer in food and occupational safety regulation in the UK. *Environment and Planning A*, 47, 373–391.

Despret, V. 2004. The body we care for: Figures of anthropo-zoo-genesis. *Body & Society*, 10, 111–134.

Environment Food and Rural Affairs Committee of the House of Commons, 2013. *Contamination of Beef Products Eighth Report of Session 2012–13*. http://www.publications.parliament.uk/pa/cm201213/cmselect/cmenvfru/946/94602.htm Accessed 30 October 2015.

European Commission 2002. Regulation (EC) No. 178/2002 of the European Parliament and of the Council of 28 January 2002, laying down the general principles and requirements of food law, establishing the European Food Safety Authority and laying down procedures in matters of food safety. *Official Journal of the European Communities L 31/1*. Brussels.

European Commission 2004a. Regulation (EC) No. 852/2004 of the European Parliament and of the Council of 29 April 2004 on the hygiene of foodstuffs. *Official Journal of the European Communities, L 226/3*, Brussels.

European Commission 2004b. Regulation (EC) No. 853/2004 of the European Parliament and of the Council of 29 April 2004, laying down specific hygiene rules for food of animal origin. *Official Journal of the European Communities L 226/22*, Brussels.

Food Standards Agency 2009. *Illegal Meat: Guidance for Local Enforcement Authorities in England*. London.

Food Standards Agency 2010b. *Meat Industry Guide: Guide to Food Hygiene and Other Regulations for the UK Meat Industry*. London https://http://www.food.gov.uk/business-industry/meat/guidehygienemeat Accessed 30 October 2015.

Food Standards Agency 2011. *Safer Food for the Nation: Food Standard's Agency's Strategy to 2015*. http://tna.europarchive.org/20141204090942/http://www.food.gov.uk/sites/default/files/multimedia/pdfs/strategy.pdf Accessed 30 October 2015.

Foucault, M. 2007. *Security, Territory, Population: Lectures at the College de France 1977–78*. London: Palgrave Macmillan.

Hampton, P. 2005. *Reducing Administrative Burdens: Effective Inspection and Enforcement*. London: HMSO.

Hinchliffe, S. 2001. Indeterminacy in-decisions – Science, policy and politics in the BSE crisis. *Transactions of the Institute of British Geographers*, 26, 182–204.

Howard, M.T. 2004. Food hygiene regulation and enforcement policy in the UK: The underlying philosophy and comparisons with occupational health and safety law. *Food Service Technology*, 4, 69–73.

Hutter, B.M. 1988. *The Reasonable Arm of the Law? The Law Enforcement Procedures of Environmental Health Officers*. Oxford: Clarendon Press.

Hutter, B.M. 2011. Understanding the new regulatory governance: Business perspectives. *Law & Policy*, 33.

Hutter, B.M. 2012. Risk regulation and food safety in the UK: Change in post-crisis environments. *Governance and Globalization Working Paper Series 33*. China: Sciences Po.

Hyde, R. 2015. *Regulating Food-borne Illness: Investigation, Control and Enforcement*. Oxford: Hart Publishing.

Jackson, P. 2015. *Anxious Appetites*. New York: Bloomsbury Academic.

Latour, B. 2004. How to talk about the body? The normative dimension of science studies. *Body & Society*, 10, 205–229.

Law, J. & Mol, A. 2008. Globalisation in practice: On the politics of boiling pigswill. *Geoforum*, 39, 133–143.

Lawrence, F. 2013. Horsemeat scandal: The essential guide. *The Guardian*.

Lemke, T. 2015. New materialisms: Foucault and the 'Government of Things'. *Theory, Culture and Society*, 32, 3–25.

Loeber, A. 2011. The food chain reforged: Novel food risk arrangements and the metamorphosis of a metaphor. *Science as Culture*, 20, 231–253.

Marsden, T., Lee, R., Flynn, A. & Thankappan, S. 2009. *The New Regulation and Governance of Food: Beyond the Food Crisis?* London: Routledge.

Miele, M. 2013. Religious slaughter: Promoting a dialogue about the welfare of animals at time of killing. *Society and Animals*, 21, 421–424.

Miyazaki, H. 2003. The temporalities of the market. *American Anthropologist*, 105, 255–265.

Mol, A. 2002. *The Body Multiple: Ontology in Medical Practice*. Durham, NC: Duke University Press.

Nancy, J.-L. 2008. *Corpus*. New York: Fordham University Press.

Pennington, H. 2009. *Public Inquiry into the September 2005 Outbreak of E. coli O157 in South Wales* http://gov.wales/topics/health/protection/communicabledisease/ecoli/?lang=en Accessed 30 October 2015.

Popper, D.E. 2007. Traceability: Tracking and privacy in the food system. *Geographical Review*, 97, 365–388.

Scottish Government 2013. *Report of the Expert Advisory Group*. http://www.gov.scot/resource/0042/00426914.pdf Accessed 30 October 2015.

Serres, M. 2008. *The Five Senses: A Philosophy of Mingled Bodies*. London: Continuum.

Star, S.L. & Griesemer, J. 1989. Institutional ecology, 'translations' and boundary objects: Amateurs and professionals in Berkeley's Museum of Vertebrate Zoology, 1907–39. *Social Studies of Science* 19, 387–420.

Tombs, S. 2015. *Social Protection After the Crisis: Regulation Without Enforcement*. London: Policy Press.

Torny, D. 1998. La traçabilité comme technique de gouvernement des hommes et des choses. *Politix*, 11, 51–75.

UK Government, 1990. *Food Safety Act, 1990*. http://www.legislation.gov.uk/ukpga/1990/16/contents Accessed 30 October 2015.

Unison 2013. *Pig Meat Inspection Health Risk to Consumers*. http://www.unison.org.uk/news/article/2013/11/pig-meat-inspection-health-risk-to-consumers/Accessed 30 October 2015.

Chapter Seven
A Surfeit of Disease: Or How to Make a Disease Public

The disease emergencies with which we started this book are labelled as such because they affect or are thought to have the potential to affect people, often in large numbers and at speed. Epidemics (so called because they involve a rate and quantity of infection above an expected background rate) and pandemics (where infections range across more than one health region or continent) are terms that instil fear concerning not only the experience of illness, but also the possible collapse of social order if so many people fall ill at the same time. Whether this fear is widely shared and indeed how people respond to it are key determinants of how a disease risk will play out. On the one hand, panic and anxiety can produce deleterious effects, while nonchalance and fatalism can also make a disease event more difficult to manage.

Infectious and food-borne diseases are, in this sense, public affairs. Collective or aggregated actions can mitigate as well as amplify risks. In simple terms, these actions may depend on how people engage with the issue. In the *Campylobacter* situation (Chapter 4) for example, there are spikes in the annual incidence of *Campylobacter* food-sickness during spells of warm weather and in the barbeque season (Nichols et al., 2012). Undercooking chicken meat, and preparing foods in the outdoors, is linked, unsurprisingly, to a heightened risk of food poisoning. In this case, it may seem obvious to inform and educate people in order that they take extra care with food preparation.

The public nature of diseases is more than this, however. For diseases and disease risks are matters around which a public may assemble, and possibly divide

Pathological Lives: Disease, Space and Biopolitics, First Edition. Steve Hinchliffe, Nick Bingham, John Allen and Simon Carter.
© 2017 John Wiley & Sons, Ltd. Published 2017 by John Wiley & Sons, Ltd.

(Latour, 2005a). How a disease public is assembled becomes critical, we would argue, in making a disease situation. A disease that instils political outrage concerning resources and expertise is rather different to one that sparks little but a vague sense that something might be done about a food safety warning.

A key question here, both in terms of the public as an aggregate of actors and as something sparked into being by an issue (Marres, 2005), is the extent to which people can or will engage with a matter of concern? One mainstream approach as we will see is to assume that people need to be informed, in order to increase uptake of vaccinations, instil better food handling practices and so on, and that there is a *deficit* in their understanding of the issues. This ABC (changing Attitudes leads to Behaviour change and better Choice) model is one that treats people as rational, decision-making individuals rather than socially constituted actors whose actions and practices are materially constrained, culturally shaped and governed by various types of competence and know-how (Shove, 2010). One aspect of this more embedded sense of action is that for many people there is a *surfeit* rather than a deficit of information that relate to any particular issue, be that health, food, finances and so on. In this chapter, we use fieldwork conducted after a series of infectious and food-borne disease episodes to suggest that, when it comes to disease risks, people are more often dealing with too much rather than too little information. How this surplus is managed becomes a critical question in terms of whether or not disease risks capture a public.

So our aim then in this chapter is to investigate how disease publics might be assembled when this surfeit is taken into account. What will it take to generate a public that is adequate to the disease emergencies with which we started this book? This chapter explores how people manage their own knowledge of infectious and particularly zoonotic disease threats in relation to scientific, public health, health education and media understandings of zoonotic disease risks.

We assume from the outset that public health like animal health is made through a patchwork of practices (Chapter 5). Rather than there being a straightforward cordon demarcating a 'safe' self on the inside and a unified danger to be faced on the outside, boundaries are being constantly made and unmade as people are connected, disconnected or partially connected with a range of threats, harms and other everyday concerns. In this sense, we take it as read that people, or certainly many people in the wealthier parts of the world, live with a continuous stream of warnings over what is healthy, what is deleterious and so on (Beck, 1992). Some of these risks will be salient, others distant and beyond the horizon of immediate concern. Some will contribute to a more generalised anxiety concerning modern life. The contribution of news media and popular culture more generally in responding to public concerns, and in helping to shift issues from background to foreground or vice versa, is clearly key in this process.

We start this chapter with a brief account of the media background to our empirical work, before outlining some of the common strategies to 'publicising disease'. Finally, we report our findings as a means to emphasise the difficulties of and requirements for raising disease publics.

The Media Background to Disease Publics

The research for this chapter was conducted following a series of high profile zoonotic and food-borne disease events. These events were widely reported in print, broadcast and social media. For example, in 2003 the viral respiratory disease known as 'severe acute respiratory syndrome' (SARS) appeared in southern China and led to around 8,000 confirmed cases (World Health Organisation, 2003). Headlines such as 'SARS out of control' (*The Mirror* 22, April 2003: 01) and 'SARS will go global warning' (*The Sun*, 15 April 2003) appeared, despite the fact that the vast majority of cases occurred in relatively few geographical regions. Soon after this there were reports of a new type of avian influenza virus, H5N1 (Chapters 3 and 8). Highly pathogenic in poultry, the virus was known to cause illness and mortality in those humans in close contact with affected birds. Although there were no instances of human-to-human transmission, there were ongoing concerns over the extremely high mortality rate (60%) amongst those infected and the possibility of viral mutations that might enhance transmissions between people. Reports on the continued spread of H5N1 infection were characterised by frequent pictures of quarantined farms, empty poultry sheds, the seeming chaos of live bird and wet markets in Asia in particular, and alarmist headlines such as 'Bird flu "plague pits" plan' (*The Sun*, 3 April 2006) and 'Bird flu: Britain braced for threat from virus' (*The Guardian*, 6 April 2006).

In 2009, reports began to appear about a novel influenza virus (H1N1) that popularly became referred to as 'swine flu' because of its believed origin in farmed pig populations. Traced back to a pig production site in Mexico, it rapidly spread amongst human populations on several continents, and on 25 April 2009, the WHO released a statement describing the outbreak as 'a public health emergency of international concern' (WHO, 2009a). On 11 June 2009, the WHO raised its pandemic alert level to Phase 6, its highest level (WHO, 2009b). The arrival of swine flu in Britain was met with intense press interest and, while many reports were restrained, others painted an apocalyptic picture of millions being infected ('Swine flu to hit millions', *Daily Mail*, 3 July 2009) and spread being out of control ('"Swine flu spreading so rapidly we can't contain it", experts say', *The Times*, 26 June 2009). Matters came to a head in July when the Chief Medical Officer warned that in a worst-case scenario the NHS should plan for up to 65,000 flu deaths ('65,000 swine-flu deaths expected', *The Times*, 17 July 2014).

These fears were never realised – the swine flu epidemic ended up as no worse than regular seasonal flu. Be this as it may, the UK Government responded with a range of measures. One of these was the creation of a *National Flu Service* that allowed sufferers access to antiviral drugs (Tamiflu) after a telephone consultation

Figure 7.1 Catch it, Bin it, Kill it campaign poster (2009). (http://webarchive.nationalarchives.gov.uk/+/www. dh.gov.uk/en/publicationsandstatistics/publications/publicationspolicyandguidance/dh_080839)

and the mass vaccination of those deemed most at risk, people with prior medical conditions, health workers, the elderly and, for the first time, pregnant women. Another measure was the launching of a large publicity campaign by the UK Department of Health to encourage behaviours to reduce the transmission of the flu virus (Hilton and Smith, 2010) under the slogan *Catch it, Bin it, Kill it* (see Figure 7.1). This involved posters, leaflets and advisements in the press and television, as well as the installation of hand sanitisers within public buildings and their commercial take off as items for personal use.

The swine-flu outbreak of 2009 was also notable because it was the first epidemic of the social media age. People used the micro-blogging service 'Twitter' and the social networking service 'Facebook' to exchange opinions and experiences about the flu epidemic. In addition, public health agencies increasingly used social media to disseminate information about the epidemic, such as encouraging the uptake of vaccination for health workers. Between May and December of 2009, over two million tweets mentioning 'H1N1' or swine flu were exchanged, with peaks during specific news events, such as celebrity infections (Chew and Eysenbach, 2010). It was also the first epidemic where it was possible for people to access 'real-time' maps of the global spread of infection (Carneiro and Mylonakis, 2009). It would appear that the era of news being 'managed' by daily press conferences, official statements and the expert 'talking head' was in rapid decline.

As well as these concerns about the spread of zoonotic infections directly from animals to people and between people, there were also fears about food-borne zoonotic infection. There have been regular reports of outbreaks associated with food-borne pathogens, both from food outlets and the locations where humans and animals interact (e.g. 'Blunders at *E. coli* petting Farm', *Daily Mail*, 16 June 2011). Despite the evidence of UK food-borne infection undergoing a decline since the early 1990s (Gormley et al., 2011), it was still the case that some outbreaks could cause significant media and policy interest. For example, in 2011 a novel and serious strain of *Escherichia coli*, originating in Germany, caused widespread illness with cases throughout Europe and sensational headlines ('1,800 victims of deadliest *E. coli* strain ever treated', *The Express*, 4 June 2011). More recently, and as we discussed in Chapter 4, policy-makers have ongoing concerns about increases in the rate of *Campylobacter* infections from domestic poultry (Nichols et al., 2012).

Disease events, such as these, have been amplified in popular film and television culture. Some, such as the Hollywood movie 'Contagion' (2011), attempted to depict a scientific and procedurally accurate portrayal of how scientists and policy-makers might respond to a deadly flu pandemic. Others avoided any attempt at realism and used narratives more firmly based in the fantastic and grotesque. Thus the movie '28 Days Later…' (2002) depicts the catastrophic accidental release of an engineered zoonotic virus. More recently the currently resurgent zombie genre, such as the AMC television series 'The Walking Dead' (2010) or the film 'World War Z' (2013), could arguably speak to a range of contemporary concerns as diverse as Alzheimer's disease or the long, drawn-out crisis in the capitalist market economy. But the collective fear of zoonotic plagues and infection also play a more obvious and immediate part in the popularity of this genre. Indeed, as Lynteris (2015) has observed, the speeding up of zombies in contemporary film reflects a possible shift from an implied critique of capitalism (creating sleep walking and living dead 'subjects') to a fear of superspreaders and the sudden collapse of social order in conditions of global interconnection.

Publicising Disease: From Public 'Understanding' to 'Engagement'

This was the background context in which we began to examine what people made of the threats posed by zoonotic diseases and how they might engage with formal and informal understandings of infectious events. Much of the public debate about the dangers posed by zoonotic disease draws on reified and disparate scientific knowledge(s) – biomedicine, epidemiology, microbiology, virology and genetics have all been brought to bear on interpreting zoonotic infections. Versions of these knowledges are then mediated and translated for social groups via broadcast and print media and government campaigns. In addition, people including journalists and other commentators can also directly access scientific outputs through open access publishing, some of which have a 'publish first judge later policy' (Robb, 2014), as well as blogging and micro-blogging sites and social media. The result is a proliferation and surfeit of information, with only some of it peer reviewed.

Questions and concerns about how people come to grips with scientific knowledge(s) have existed since The Enlightenment (Irwin, 1995). Contemporary discussion of science and its publics often refers back to The Bodmer Report published by the UK's Royal Society entitled the *Public Understanding of Science* (Royal Society, 1985). This stressed the economic, social, political and cultural benefits that an increase in citizens' scientific literacy would bring to the nation. As such the report can be seen as an attempt to align various actors, including lay publics, the media, scientists and policy-makers to the process of facilitating a general understanding *of* science *by* the public (Michael, 1996). The report was however heavily criticised for its implicit *deficit model* contained in the very idea of the 'public understanding of science' (Wynne, 1995). The implicit assumption was that where once there was a lack, 'correct' knowledge could be acquired through citizens consuming a 'black-boxed' science that would then allow them to behave 'correctly', rather than 'irrationally' when making decisions.

The model of the 'public' as suffering from a 'lack' causing irrational behaviour underwent a dramatic shift due to an earlier zoonotic crisis. In 1996, after years of official denial, the UK Government announced that there was a probable link between bovine spongiform encephalopathy (BSE) and human Creutzfeldt–Jakob disease (vCJD) and that human consumption of beef was the likely cause of vCJD (Hinchliffe, 2001, 2007). The *volte face* and the events leading up to the admission generated what many saw as a new low in terms of public trust in science and were seen as a watershed in terms of publics, risk and science. As a result, a post-BSE consensus emerged that it was no longer justifiable to mistrust the public with scientific information and there was a need for public discussion of uncertainties regarding risks. This discussion and any resulting decision-making should be open, transparent and accountable (Jasanoff, 1997; Phillips et al., 2000). Government reports began, in the early 2000s, to stress how a 'crisis of trust' required a new era of dialogue between scientists, experts, policy-makers and lay publics (House of Lords Select Committee on Science and Technology, 2000) and the modality, at least in principle if not in practice, switched from 'pubic *understanding* of' to 'public *engagement* with' science.

This apparent shift from understanding to engagement is indeed positive, but a tendency remains to treat science as pre-eminent, and engagement as an activity that occurs 'after the fact', as if it is only the finished products of science that will be 'engaged' by the social groups. Any resulting knowledge gap or difference between scientific opinion and the public at large tends to be understood as a problem for the 'public' rather than the science. The antidote to the implied deficit or gap is often taken to be better engagement with complex knowledge, rather than, as we would see it, the development of forms of knowledge practice that are cognisant of the complexities that are involved in the day-to-day practices of health and disease management. More radically, seen in this vein, engagement can be a more generative co-production of knowledge (Born and Barry, 2010; Hinchliffe et al., 2014; Irwin, 2006; for a parallel approach in flood management see Lane et al., 2011).

Engaging and understanding publics (rather than correcting public understanding) and taking the time to research people's day-to-day means of dealing with disease and other risks immediately prompts a shift away from any deficit approach to public knowledge. Indeed, and instead, people are understood to arrive at understandings, and have concerns, that hail from many different domains. These include multiple scientific discourses, policy discourses and popular culture (Irwin and Michael, 2003). Rather than a deficit of knowledge, there is often a *surfeit* of fragmented and heterogeneous materials and ideas. In addition, and as we noted in previous chapters, within any disease situation there is more than one disease diagram in play. The ways in which these diagrams are patched together, made sense of, ignored or otherwise, is thus a key resource for researchers interested in public engagement.

In attempting to engage the public, it is also, of course, important to ask who or what is 'the public' of which we speak. In the context of zoonotic diseases, the obvious answer might start with the concept of 'public health', even if this construct is highly contested and has multiple meanings (MacDonald, 2013). A relatively uncontroversial starting definition would be that provided by Acheson (1988: 1), who described public health as 'the science and art of preventing disease, prolonging life and promoting health through the organised efforts of society.'

The 'science and arts' of public health, even in this short statement, seems to indicate a mixture of efforts that are required to prevent acute maladies (like avian influenza) as well as respond to chronic issues (like obesity or heart disease). Rather in the manner of public understanding and public engagement arguments, this can also go two ways (Verweij and Dawson, 2007). First, there is the more deficit-like, or downstream idea of 'the public as a social entity or target for an intervention' (Dawson, 2011: 3). Here science leads, and the targeted public needs to follow by adopting advice and treatment. Second, there is a more collective notion of a public as a 'mode of intervention, which requires some form of collective action' (Dawson, 2011: 3). Here, there may be a sense in which a public is entangled with and so forms around issues. In the process, this disease public generates more or less healthy outcomes.

The target approach is beset by a number of difficulties that include but are not reducible to the deficit approach. First, there is the issue of heterogeneity. There is always more than one risk category or group. Different life stages, cultures, genders, habits and practices tend to segment the public. Second, there is the well-known paradox whereby sizable health benefits for a society of, say, achieving high rates of immunisation, offers relatively little benefit to an individual, whose chances of acquiring the disease may be quite low. Any 'measure that brings large benefits to the community but offers little to each participating individual' (Rose, 1981: 1850) has been termed the prevention paradox. The point here is that uptake of public health (or, indeed, animal health, see Enticott, 2008) measures may be lower than the social optimal, not because of public misunderstanding or a misreading of the risk, but because of more or less rational decision-making on the part of people who see relatively little personal benefit. Rationality in this sense is neither present nor absent but folded into a range of often-conflicting messages.

In this situation, a will on the part of public health experts to simplify health information may well lead to problems later on. For example, during the 'swine flu' epidemic, antiviral drugs were offered to infected people, even though evidence indicated they were of marginal benefit individually and potentially had high costs in the form of significant side effects (Dobson et al., 2015). But part of the reason for antiviral use was because it was believed that they might reduce viral shedding and thus lower rates of onward infection. The social good of taking the medicine was not it seemed matched by a personal gain in swallowing the pills. The subsequent revelations that the medicines were of questionable utility, alongside the public exposure of lucrative public-private purchase contracts, the financial costs of which were borne by the taxpayer, only served to undermine expert health advice.

A more recent approach taken by governments to constitute a 'healthy public' has switched from attempts to address deficits through 'health understandings' towards efforts to change health behaviours. This change in focus is based on 'nudge theory' (Thaler and Sunstein, 2008), which argues that suggestion and positive reinforcements (or altering the choice architecture) can be used to adapt or 'nudge' behaviour by playing on 'assets' that people already value. Here assets may include existing beliefs, economic interest or simple convenience (e.g. putting fruit at eye level in self-service food outlets). The concept is itself openly paradoxical, being based on the idea of 'libertarian paternalism': people need help to live healthy lives (paternalism), but such help should be not forced or meddlesome (libertarian). The focus is on designing social environments that make 'good' choices easier for individuals and collectives. The debate on this behaviour change approach is beyond our remit here, but suffice to say that it is attractive to policy-makers as it marks a way to alter activities without the expense or nuisance of first having to change people's attitudes. Slight shifts in choice architecture or financial incentive can, it is argued, do the job of securing already established goals.

In situations where these goals are widely shared and uncontroversial this may be a useful approach. So, for example, most people agree that more vital organs are needed for transplant operations and so there tends to be reasonable if not a general consensus that shifting the choice architecture to 'opting out' of donation rather than the more onerous 'opting in' is a good thing. However, where there is ambivalence or complexity regarding means and ends, then the approach may end up adding to an already crowded information environment and so repeat the failures of knowledge-gap or deficit approaches (Selinger and Whyte, 2012). In more complex situations there remains a pressing requirement to understand and engage publics prior to any intervention of this or other kind. It is this research need that informs the remainder of this chapter.

Understanding and Engaging Disease Publics

Against this background, we sought to investigate how people understood emerging disease threats, how they used information and how this became part of their daily routines and lives. We staged a series of ten focus groups with respondents drawn from across the UK (Scottish Highlands, Midlands, East of England, West Country, South East England and London). The purpose was to account for and avoid any metropolitan or other bias in the discussions. The groups were themed by both interests and social grouping (e.g. bird watchers, organic farmers and allotment owners, A-level students, young and mature professionals, unemployed and pensioners) and were demographically diverse (e.g. low and middle income, different ages, genders and ethnic identifications). Again, the idea was to access broad discourses regarding emerging disease from a variety of starting points. Respondents were recruited from pre-existing social groups and to a large extent already knew one another.

The method allowed us to access the complex and dynamic nature of respondents' views and attitudes. In other words, we were concerned to take advantage of the group interactions that would exist amongst people who knew each other by encouraging the research participants to talk freely to one another as well as to the researcher facilitating discussion. Additionally and most importantly in this context, the focus group methodology is particularly suited to an exploration of views that may not be fully formed and can offer us insights into how people manage uncertainty and make decisions. In other words, focus group methodology can be used to follow social processes, albeit under relatively contrived circumstances, whereby respondents 'make their minds up' and reach an opinion about an issue which they might not have considered before (Carter and Henderson, 2005). In this respect, the method can access social processes that surveys and interviews may miss or only cover superficially. They allow exploration of issues in people's own language or modes of understanding, and provide a platform wherein they can generate their own questions and pursue their own priorities (Barbour and Kitzinger, 1998; Kitzinger, 1994).

The design of the focus groups took place after other work in this broad project, which encompassed work related in previous chapters, had already started. We used activities to facilitate group interaction and discussion. One of these used initial findings from our exploration of the literature and fieldwork on farms and in the food industry to prepare a series of around 20 'trigger' statements. These statements were intentionally provocative and at the time current, and likely to cause some debate, for example: 'it's only people who have direct close contact with infected animals that need to worry'; 'governments and international organisations are able to manage new types of disease'; and 'to prevent the spread of flu catch it in a tissue, bin it and kill it by washing your hands.' Respondents in groups were asked to discuss and arrange these statements, individually written on cards, along a continuum from agree strongly to disagree strongly. The analytical focus of this activity was not to quantify agreement or otherwise with responses but rather to access group discussions, disagreements and negotiations about statements.

Another exercise involved a collection of around 25 images (e.g. scientist in a lab coat, an image of a wet market, Tamiflu, a vial of vaccine and a person in safety clothing collecting a chicken). Respondents were asked to produce their own 'news story' that they might be expected to see, hear or read either nationally or locally. After this activity, there was a final section with a set of more general open-ended questions: What would you like to see in the media? Where would you get information if you were worried? What type of information would you need to make your mind up about the danger of diseases that come from animals?

In what follows, we present analyses of the activities and discussions with the various social groups identified above. Each group was tape-recorded and transcribed, and transcripts were coded using a variety of emergent conceptual themes. Analysis involved, at the practical level, close scrutiny of transcripts and a focus not only on what was said but also how it was expressed (what form of words were used to generate understanding and so on). In that sense, our main analytic tool was discourse analysis, or the understanding that participants' utterances created meaning within the broader context of available language and sense-making. In this understanding, the discussions were both an expression of people's ability to make sense of issues and of the broader cultural processes of which they were and are active participants.

Understanding the Surfeit

In the following, we distil our findings into two broad areas. The first reports on a sense of both disgust regarding issues of contamination, but also a form of nonchalance regarding infectious and food-borne disease. In the second we expand on the important issue of *non-knowledge* or ignorance as a *resource* that people actively muster in conditions of surfeit, before drawing the chapter to a close by revisiting the issue of what kind of disease public is sufficient to the current 'emergency'.

Making Distance: From Disgust to Nonchalance

One of the first issues that we noticed during the exchanges in the groups was how people attempted to manage and make sense of the many uncertainties around zoonotic infection. We were particularly interested in how people managed issues associated with purchasing and preparing food, attempting to avoid infection, and dealing with potential infection in their everyday life and their strategies for living with microbes. It was particularly important to examine this in light of the information they received, from the diverse range of sources outlined above, about infections and both 'imagined' and real epidemics. The people who took part in these groups were all well aware of concerns expressed by Government and in the mass media about recent zoonotic infections, but did appear to be relatively nonchalant and pragmatic about their own personal threat.

For example, the groups readily recognised the slogan and materials associated with the *Catch it, Bin it, Kill it* campaign. However, it was discussed in tones that suggested the message was rather obvious, patronising and/or comical. When a group of young professionals were presented with a card with the words 'Catch it, Bin it, Kill it' printed on it, they immediately recognised the statement and animatedly began discussing the campaign. They found it laughable and made a link to a comical *YouTube* rap video ('Swine Flu SKANK'), and doubted the extent to which it would change behaviour, suggesting in fact that people were already in a sense divided between those who did and those who did not behave accordingly:

PAUL: *Everyone knew about it ['Catch it, Bin it, Kill it'] and it did become a thing, but it became a comical thing.*
SARAH: *Exactly... It was just overused...*
PAUL: *Yeah, it was quite funny.*
ALISON: *I think people who did it before they heard that did it, and I don't think it changed anything... I think whoever was going to 'Catch it, bin it, kill it,' did it before they even heard of that and it didn't change anything.*
(Young professionals, North London)

Other groups expressed their revulsion towards coughing and sneezing on public transport in particular, and for them, the campaign served to reinforce ideas of good manners and civility. For example, one man told a story about an encounter he had on the way to the focus group, with others nodding in agreement:

FELIX: *It's good to raise awareness because I mean people were coughing all about the place without putting their hands over their mouth.*

KIRA: *Yeah.*

DONNIE: *It's a pet hate of mine.*

FELIX: *It's a pet hate of mine. I had to tell a man today about it… a man was coughing, I said, 'Put your hand over your mouth, I don't want to breathe in your things'…*

DONNIE: *Quite right.*
             (Unemployed Group, South London)

Disgust and fear of contagion was a real and visceral concern. People were in that sense already primed regarding the sociality of infectious diseases and perhaps did not need much encouragement. And yet, beyond this immediate reaction to the bodies and emissions of others, which tap into a deep-seated, historical memory of contamination and communicable diseases (Chapter 3), there was a rather pragmatic and even fatalistic sense that there was little that could be done to shield oneself from social and microbial entanglements. This apparent mismatch between a sense of civility and pragmatic acceptance begs the question of how people juggle multiple disease diagrams. When does risk aversion turn to fatalism, and how does the alarm concerning shared spaces and bodily fluids become a mere shrug in the face of warnings over pandemics?

It has long been understood that people are rather good at balancing information from different sources about risk together with their own personal decision-making relating to everyday activities (Green et al., 2005; Kerr et al., 2007; Wynne, 1992). Our findings resonate with much of this research. In discussion, participants actively weighed up different sources of information about disease and health in order to judge what action, if any, to take. But they also balanced issues to do with their own personal health with the range of information that was available.

For example, there was a general distrust around many of the messages received about zoonotic infection. Mediated risk and disease were viewed with a high degree of suspicion. Panics and epidemics sold papers and so should not necessarily be taken too seriously:

JUNE: *I'm not saying they are panics [newspapers], I think they love them though [epidemics]… They love a good disease.*

POLLY: *They certainly crank it up, don't they?*

SUZI: *They do.*
             (Professionals, South Midlands)

Clearly, food-borne and infectious diseases can sell papers, and people like to read about possible emergencies, rather as many like to hear about impending storms or bad weather. Even so, the distance conveyed by third person plurals ('They') suggests that the frisson of emergency may remain in print only. The degree to which people feel either the need to or benefit from acting on this news may well be rather small.

In relation to this, and beyond media focus, others cited the frequent and over-cautious nature of health warnings. Warnings that originated from experts and government about the potential for avian flu and the anticipated scale of human infection during the swine flu epidemic prompted some mirth and reasons to be sceptical. As we observed earlier, the direst of warnings at the beginning of the swine flu epidemic, of possible death in the tens of thousands, were not realised. While discussing the way in which the public health campaign during swine flu was handled, this student made the following observation:

> *I had a disbelief in the whole thing really from the beginning, after the whole bird flu thing which never spread and they made it sound like it could be the end of the human race and it never even happened. So few people got it, and then also swine flu. In fact I heard that quite a few times that normal flu was actually worse, more dangerous than swine flu, yet that's not made out, it was made out that just because it came from an animal, it was so horrendous.* (A-level student group, East Coast England)

In some senses, the apparent acceptance of contamination messaging was being tempered by a sense that not all infections, or disease events, required alarm. Some were less dangerous than others. It depended on so much more than simply being infected. Configurational approaches to disease (Chapter 3) were familiar currency in people's discussions.

Despite the sensationalism and a tendency in the media to fear the new, people were likely to treat disease risks with a certain amount of caution and even disdain. Again, there was a distance between news stories and people's felt need to do much more than register them as stories. This distance was not simply there – it was actively made in conversation. Towards the end of the focus groups, for example, we asked what type of information would be useful to allow people to make judgements about some of the issues that had been discussed:

> *If you're bombarded with things you should be scared of and then, 'But oh no, maybe I should be scared about this instead'… but I think… getting bombarded with all this different information and not knowing your way through… I just think we don't actually… a lot of it is actually really not going to affect our lives and isn't really relevant to us…* (Organic Gardeners Group, South West England)

'Bombardment' speaks volumes, we would suggest, of a surfeit of information, and requires, it seems, active filtering in order to prioritise certain 'relevant' issues over others. People were not in this sense lacking information, but actively choosing to ignore some issues, or to give them a lower priority.

In order to do this, people drew on their own experience of infectious diseases (colds, influenza, measles, chicken pox and so on) and were as a result sceptical regarding 'simple models' of contamination, containment and closure. Indeed our findings suggest that people understood health in broader terms than simple 'sanitation' or the avoidance of contamination and had of their own accord problematised simple 'closure models' or exclusion diagrams of disease.

Part of the issue here was that people were sceptical that closure of a boundary, whether it belonged to a nation state or to a body, was even possible. A kind of fatalism resulted which tended to make people even more likely to ignore over-zealous injunctions to decontaminate living space. One of the statement cards we asked people to discuss said 'Tighter border controls could help stop animal diseases from spreading into Britain.' This prompt resulted in the following exchanges where the idea of 'closure' on a state level was regarded as unachievable:

CLARRIE: *Yeah, I was thinking of borders being on a map, and air goes over borders... I was originally thinking about bird flu and things like that, because you can't border control the sky...*

MODERATOR: *How does everyone else feel about that?*

EDDIE: *I don't think it would make any difference.*

MODERATOR: *Why don't you think it makes a difference?*

EDDIE: *Well very much for the same reasons, you could control some borders, you can't control all borders. So if you can't control everything it's not worth the paper it's written on.*

WILL: *You can't control the birds, can you?...*

(Low Income Group, Rural South West England)

This failure of control was a common feature of discussions. Spill-over of disease and indeed spill-back, or mixtures between populations of wild, domestic and human animals seemed to be something that people expected. Indeed, if anything, modern farming seemed to be the problem rather than the solution:

> It's always going to happen [spread between wild and domestic animals], and the difference is that domestic animals are given things like... Jabs, antivirals, anti-bacterials to get rid of those diseases. When they then contact wild animals, there's transfers both ways of the same diseases but because they are domestic animals they've been given the antibiotics and stuff, they then get resistant versions of the same disease that then gets transferred back into the wildlife. (Unemployed Group, South London)

This paradox of control, or control leading to further problems later on, was often supported by an appeal to the benefits of exposure to microorganisms.

While sneezing on the bus provoked disgust, there were also countervailing stories to tell regarding building up immunity through experience and a concomitant lamentation of the decline of outdoor life and (for children) play. In ways that resonate with the pig farmers we met in Chapter 5, people drew on arguments regarding the importance of building immunity through environmental exposure and microbial challenges:

> When we were children we played in muck, we built up an immune system... I mean I'm [Highland City] born and bred and there were a lot of youngsters in the street and we would go a way up to the creek or the farm for the day and play there and maybe take a piece [sandwich] with us and if we were caught short you would go behind a bush. You don't wash your hands. And we made mud pies around the back and things like that. We built up an immunity. (Pensioner Group, Highlands of Scotland)

Making Ignorance in Conditions of Surfeit

Mistrust of modern institutions (including farming) is unsurprising, but the broader question for us is how do people make sense of competing ideas about health and disease when there is a *surfeit of knowledge* and *a deficit of authority*? A key issue here was the ability to employ distinct anti-epistemological discourses of ignorance and non-knowledge as an active choice and strategic resource. As we have pointed out, public uncertainty and ignorance are often seen by policymakers as voids which need to be filled. Indeed, in the fields of public health and health education, passive ignorance is often equated with death. For example, the 1980s UK Governments 'Don't Die of Ignorance' campaign to combat spread of the HIV virus is an obvious example of how risk behaviour was tied to simple notions of knowing the 'facts' (Carter, 1995). Subsequent campaigns, while more subtle, have used similar messages (e.g. campaigns around vaccination and cancer screening) to try and alter behaviour.

Ignorance or other forms of non-knowledge (uncertainty, ambivalence and so on) may well be problematic in many areas of public health, where there is a vital need for people to act on relatively straightforward measures to avoid dangerous infection or contribute to a collective immunisation. However, commonly, ignorance and non-knowledge co-exist with knowledge. Here we are not pointing towards a mere lack of understanding arising from passivity and failure to grasp the 'facts' that would be suggested by the 'deficit model' or the semi-conscious decision-making implied by 'nudge theory'. Non-knowledge here has a positive aspect 'as the twin and not the opposite of knowledge' (McGoey, 2012: 3). It can, for example, include the proliferation of knowledge around an issue as a means to actively confuse or diffuse a situation; an example would be the role of tobacco companies and lobbyists in the history of smoking and public health, or more recently the actions of motor car manufacturers in the dissimulation of pollution data. It can also refer to the purposeful development of degrees of ignorance, or of ignoring issues, as a means to cope with complexity or anxiety. In terms of the latter, sociologist Mike Michael illustrated how discourses of ignorance are used by people to 'reflect on, and articulate, their social relationships to science and its institutions' (Michael, 1996: 122). Thrift (1996) has written of the intrinsic role of social unknowing in any given society. More recently, McGoey (2012) has usefully highlighted three modes of the strategic uses of ignorance as knowledge: ignorance as emancipation, where ambiguity or 'performative silence' is used as a rebuke to classificatory systems and dogmatic certainties; ignorance as a commodity, where uncertainties over the exact status of knowledge (e.g. tobacco and health, climate change) are used for commercial advantage; and ignorance as pedagogy, where it can be used as a weapon of usurpation 'resistant both to mastery and to monopolisation' (McGoey, 2012: 10).

While McGoey has mainly focused on the strategic deployment of ignorance by organisations and groups, we are here concerned with how ignorance is strategically marshalled by people in their day-to-day lives. Far from being a failure or gap, ignorance might be understood here as one of the assets that people are able to draw on to make sense of the overabundance of often conflicting information on topics and issues, that may potentially interest them, but to which for a variety of reasons they cannot attend to. The observation is not new. Urban life, Georg Simmel observed, brought with it an intensification of sensual stimulants and this required that people find a way to soften these perpetually shifting stimuli by forming a 'protective organ' – 'There is perhaps no psychic phenomenon which is so unconditionally reserved to the city as the blasé outlook' (Simmel, 2010 (1903): 105). This *blasé mindset* is a rational and matter-of-fact response to ceaseless assault on the intellect. We would follow this to broadly characterise respondents as using a form of *blasé ignorance* – one that, as we shall see, was entered into with a full and mindful awareness and often demonstrated an underlying sophistication in dealing with disease.

We have already seen how respondents comfortably talked about many aspects of zoonotic diseases. Indeed they were often knowledgeable and appeared to be able to balance information they received from a variety of sources within their own lives and activities. However, in parallel with this, there was a real sense in which people had limits on the amount of information they were prepared to engage regarding disease and health issues. A common theme here was the way in which participants marshalled blasé notions of ignorance in their accounts of risk. They did this in several ways. One of the first was avoidance of information that was seen to be of little value. Here ignorance was summoned as a strategy for distributing attention to other more pressing concerns:

> I don't watch the news anymore and I haven't died [laughter], you know, so I haven't keeled over. (Organic Gardener, South West England)

There is cynicism here but also a shared humour regarding the fact that ignorance might be the only really healthy way of existing in an overly-reflexive world (the classic account is Beck, 1992). Indeed, knowledge was seen as flawed and egregious and may sometimes only lead to anxiety. People in this sense were all too aware of the pernicious nature of health knowledge. As we noted in Chapter 5, for Martin (1995: 122), this is the 'paradox of feeling responsible for everything and powerless at the same time, a kind of empowered powerlessness.'

Ignoring is a way of dealing with a surfeit of information, but it is also a means of bracketing out the unknowns over which people had little personal control. For example, in a discussion of urban life, and particularly the day-to-day use of an overcrowded urban transport system, the potential stresses of being concerned about every possible source of infection prompted a form of coping that involved actively bracketing out any possible risks:

> *There is a bit of 'ignorance is bliss', in that if there's nothing that I can actually do to protect myself, I don't want to be every day worrying about things, if there's nothing that I can do.*
> (Young Professional Group, North London)

Not only were these known risks made into unknowns by bracketing them out, or refusing to think about them, ignorance was also strategically used by respondents as a form of dynamic mindfulness that further knowledge may be counter-productive. One expression of this was the idea that 'knowledge' may make daily decision-making impossible. This was most easily expressed in the context of food safety:

> *Do you worry about something being organic? Do you worry about something being in plastic? Do you worry about it being fair trade? You know, do you worry about the health issues and so for me, that's part of the reason that with health concerns, that if it doesn't really affect me as a healthy middle-aged woman and not pregnant and not old and not a child, I just think 'I'll let that one go', because there's just too many other things.*
> (Professional, South Midlands)

Here the risk of becoming obsessed, or compulsive, was seen as a bigger danger than taking a more fatalistic approach to the possibility of communicable diseases (on anxiety and food consumption practices see Jackson, 2015; on swine flu anxiety see Everts, 2013). Making oneself immune was all very well as a means to live well, but it had better not turn into an 'auto-immune' disease where constant vigilance turns against itself:

BILL: *There's not much you can do about it on a day-to-day basis. I mean you can be aware that it exists and you just have to hope that...*

ERIC: *That it doesn't come to you.*

MAU: *I think you just get on with your everyday life...*

MAU: *You can't have it there every minute of the day because you would become obsessed. You could go mad. You've got to live your life.*
(Pensioners, Highland Scotland).

Ignorance was in this sense an active strategy that people take in order to carry on; to 'keep calm and carry on', became a somewhat over-used slogan in the early part of the twenty-first century. It keyed into a nostalgia for a certain kind of national character associated with the 'real' tests that people faced in the Second World War, but its somewhat ironic re-emergence might also say something about the bio-, climate and terror emergencies of the current period.

As we have tried to suggest, ignorance was a strategy that involved sorting knowledge rather than necessarily rejecting it altogether. In this sense, avoiding information overload was handled in other ways by choosing only to focus on materials that related to immediate interests, or to use trusted sources. Few if any sources of information were seen as neutral, whether they were official or more obviously related to specific interests. So a typical conversation would see a participant cite an outside 'authority', only to have that authority quickly doubted by others. The end result of the surfeit of information coupled with a deficit of authority was characteristically, for some members of the groups, a decision to revert once more to ignoring everything:

PETER: *I don't really eat meat anymore. I occasionally eat meat but I try not to eat red meat and stuff.*

WILL: *I always just wonder how much evidence there is behind it. Like you hear so many things but is it actually true?... I don't think it's true to be honest...*

PETER: *There is evidence, and organisations like PETA [People for the Ethical Treatment of Animals] have evidence.*

KALINDA: *It does sometimes feel like there's so many things that you could potentially worry about with food and contamination and [flu] risk, it's almost like an overload. So I think somewhere I've just gone, 'Right, there's no way I can process all of this. I'm almost going to try and ignore all of it.'*
(Young Professional Group, North London)

In our focus groups, authority had taken something of a 'hit' following the swine flu 'pandemic', which was regarded as a false alarm by participants. Following a discussion about a statement card which read 'governments and international organisations are able to manage new types of disease,' there was a consensus that this was not the case. A distinction was drawn between 'hope' (that disease would be managed by governments) against 'experience' (that they would be unable to do this), a scepticism fuelled by swine flu but also foot and mouth and the lasting memory of burning pyres of animals in the British countryside. One discussion came to a head when a woman connected the idea that 'swine flu never happened', at least to the anticipated extent of some predictions made by government experts, with the financial benefits to be had by commercial concerns during the epidemic:

The swine flu one was supposed to be a pandemic and... really, really serious and it just fizzled out, and huge amounts of money were spent on it. They didn't handle that very well... it didn't give a lot of confidence because they didn't handle it very well. (Low Income Group, South West England)

Spending huge amounts of money on anti-viral drugs, in particular, had been in the news, and people expressed some concerns that they were being asked to sanction public spending on the basis of what they regarded as highly questionable knowledge claims. These kinds of connections produced lasting damage to authority. The result was, as we have already suggested, a heady concoction of a surfeit of information and a deficit of authority. Given that disease emergencies require collective action, the question remains how this impasse might be overcome?

Conclusions: Making a Disease Public

In focus groups, people tend to be quite nonchalant and pragmatic about the threats that relate to disease emergencies, and particularly those that relate to zoonotic disease. The people who took part in these groups refused the simple 'cordon sanitaire' model of biosecurity with a safe inside, risky exterior and hyper-vigilant observation of borders. They did this by drawing on experience, a sceptical attitude towards authority and expertise and a form of strategic ignorance. This is not to say that they were unconcerned about contagion. Indeed many expressed entirely reasonable wishes about avoiding infection and the maintenance of their own good health. They expressed concerns about microbes and pathogens while acknowledging that the locus of control was often elsewhere and that their own knowledge and that of others was riven with uncertainty.

Indeed, pandemics, zoonoses, food-borne illnesses and expert knowledge were all associated with uncertainties, something that made any simple equation between information and action problematic. Uncertainty is frequently discussed in the social sciences but often treated as a unified or opaque concept. As Christley et al. (2013) usefully point out, there are differing forms of uncertainty. There are *weak* uncertainties that can be expressed in terms of probabilities; *strong* uncertainties where the range of outcomes may be known but where probabilistic reasoning cannot be assigned; *epistemic* uncertainties where unknowns can be resolved through further research; and *ontological* uncertainties where outcomes are unknowable because causal chains are open or highly contingent. In making these distinctions the authors are specifically talking about the modelling of epidemics, but here, in the realms of living with disease threats, we could add a fifth kind of uncertainty. In a situation of multiple sources of information, with a chronic questioning of authority, people seem to live with a form of social uncertainty that is quickly turned into a form of social security through the active production of ignorance or non-knowledge. In other words, the surfeit of information about risk led to a willed and strategic ignorance as a coping strategy.

This production of 'ontological security' (Giddens, 1991) through non-knowledge is highly rational in what Martin (1995) described as the paradox of being made to feel responsible for something (health) which is to a greater or lesser extent beyond people's power of control. The continual vigilance associated with a neoliberal subjectivity is resisted in that sense through appeals to everyday experience of health and illness (and keeping well, as well as sane) and through the active filtering of knowledge. This matters, first, because it militates against a policy of simply informing the public in order to achieve action and compliance. Adding yet more information into the social realm may not always produce effective responses in conditions of surfeit. To repeat, there is no deficit that needs to be filled. Rather there is a set of resources and understandings that are often carefully assembled by people from experience. Second, we take it as read that any effort to address emerging disease risks will need to engage people in those risks. It is clear that risks can be amplified or indeed reduced depending on how people respond to an emerging threat. For example, as we showed in Chapter 4, because of the way that poultry is produced, Campylobacterosis is a constant risk for consumers, and the extent of this risk will depend among other things on kitchen practices. Any attempt to engage people in reducing risk will need, in our view, not only to present this information (poultry needs to be well-cooked, preparation surfaces need to be kept clean, poultry should not be washed under a running tap and so on), but also accept that information does not enter a void – it is made sense of and possibly actively ignored in ways that are responsive to experience, trust in authority and a surfeit of information. This is not a recipe for giving up on public health, only a plea to move on from a deficit model and seek detailed social science understanding of how people actively deal with risk.

The surfeit and the resulting coping strategies mean that we can and should go further, and this involves thinking again about risky or disease publics. Earlier in this chapter we asked who the public of 'public health' was. We argued that this public is often constituted as part of an issue-led response to a crisis – the public becomes both a social target and a mode of intervention. We have seen how this sometimes becomes quite problematic. The people who took part in these groups often failed to recognise themselves as a target, and were not interpellated by messages about zoonotic threats from public health, health education and the media about disease risks. Alternatively, when people did sometimes accept that they may be at risk from some form of zoonotic threat they often felt 'bombarded with messages' and retreated into a *knowing* but blasé ignorance. This raises the prospect that it may be becoming increasingly difficult to actually raise a public for public health initiatives through targeting. Making a disease public is clearly a matter that needs to recognise that any such public will be a matter of division, contest and debate.

The issue here may be less palatable than certain readings of nudge theory and public understanding of science. The pathological lives that are to be found in this book may require a rather different emergency to be addressed than the one that tries to educate people about dangerous pathogens. This is a disease public that seeks to address the unpalatable conclusions that open up pathogenicity as a question. Regularly buying a commercial product that is 'toxic' or potentially 'fatal' may be a more salient issue for public health than pleas for proper food preparation. The latter attempts to delegate responsibility for safety from a few commercial spaces (e.g. food producers and supermarkets) to millions of domestic locations. It may be that in this case the rather passive construction of consumers as those who can be nudged into better food preparation or disease avoidance is insufficient to the current crisis or emergency. For Timothy Lang, the *Campylobacter* situation is a public health scandal rather than a public health campaign (Lang, 2014). We take it that what is meant here is the need to raise the profile of the extent of pathological lives, to make them political (or cosmopolitical – see Chapters 8 and 9), and only in that way can a truly public health be raised.

References

Acheson, D. 1988. *Public Health in England*. London: HMSO.

Barbour, R. & Kitzinger, J. 1998. *Developing Focus Group Research: Politics, Theory and Practice*. Thousand Oaks, UK: Sage.

Beck, U. 1992. *Risk Society*. London: Sage.

Born, G. & Barry, A. 2010. Art-science: From public understanding to public experiment. *Journal of Cultural Economy*, 3, 103–119.

Carneiro, H.A. & Mylonakis, E. 2009. Google trends: A web-based tool for real-time surveillance of disease outbreaks. *Clinical Infectious Diseases*, 49, 1557–1564.

Carter, S. 1995. Boundaries of danger and uncertainty: An analysis of the technological culture of 'risk assessment'. *In*: Gabe, J. (ed.), *Medicine, Health and Risk*. Sociology of Health and Illness: Monograph Series. UK: Wiley-Blackwell.

Carter, S. & Henderson, L. 2005. Approaches to qualitative data collection in social science. *In*: Bowling, A. & Ebrahim, S. (eds), *Handbook of Health Research Methods*. Maidenhead: Open University Press.

Chew, C. & Eysenbach, G. 2010. Pandemics in the age of Twitter: Content analysis of Tweets during the 2009 H1N1 outbreak. *PLoS ONE*, 5. e14118. doi:10.1371/journal.pone.0014118

Christley, R.M., Mort, M., Wynne, B., Wastling, J.M., Heathwaite, A.L., Rickup, R., Austin, Z. & Latham, S.L. 2013. 'Wrong, but useful': Negotiating uncertainty in infectious disease modelling. *PLoS ONE*, 8. e76277. doi:10.1371/journal.pone.0076277

Dawson, A. 2011. Resetting the parameters: Public health as the foundation for public health ethics. *In*: Dawson, A. (ed.), *Public Health Ethics: Key Concepts and Issues in Policy and Practice*. New York: Cambridge University Press.

Dobson, J., Whitley, R., Pocock, S. & Monto, A. 2015. (2015) Oseltamivir treatment for influenza in adults: A meta-analysis of randomised controlled trials. *The Lancet*, 385, 9979, 1729–1737.

Enticott, G. 2008. The ecological paradox: Social and natural consequences of the geographies of animal health promotion. *Transactions of the Institute of British Geographers*, 33, 433–446.

Everts, J. 2013. Announcing swine flu and the interpretation of pandemic anxiety. *Antipode*, 45, 809–825.

Giddens, A. 1991. *Modernity and Self-Identity. Self and Society in the Late Modern Age.* Cambridge: Polity Press.

Gormley, F.J., Little, C.L., Gillespie, I.A., Lebaigue, S. & Adak, G.K.E. 2011. A 17-year review of foodborne outbreaks: Describing the continuing decline in England and Wales (1992–2008). *Epidemiology and Infection*, 139, 688–699.

Green, J., Draper, A., Dowler, E., Fele, G., Hagenhoff, V., Rusanen, M. & Rusanen, T. 2005. Public understanding of food risks in four European countries: A qualitative study. *European Journal of Public Health*, 15, 523–527.

Hilton, S. & Smith, E. 2010. Public views of the UK media and government reaction to the 2009 swine flu pandemic. *BMC Public Health*, 10, 697.

Hinchliffe, S. 2001. Indeterminacy in-decisions – Science, policy and politics in the BSE crisis. *Transactions of the Institute of British Geographers*, 26, 182–204.

Hinchliffe, S. 2007. *Geographies of Nature: Societies, Environments, Ecologies.* London: Sage.

Hinchliffe, S., Levidow, L. & Oreszczyn, S. 2014. Engaging cooperative research. *Environment and Planning A*, 46, 2080–2094.

House of Lords Select Committee on Science and Technology 2000. *Third Report: Science and Society*, HL Paper 38. London.

Irwin, A. 1995. *Citizen Science: A Study of People, Expertise and Sustainable Development.* London: Routledge.

Irwin, A. 2006. The politics of talk: Coming to terms with the 'new' scientific governance. *Social Studies of Science*, 36, 299–320.

Irwin, A. & Michael, M. 2003. *Science, Social Theory and Public Knowledge.* Maidenhead, UK: Open University Press.

Jackson, P. 2015. *Anxious Appetites.* New York: Bloomsbury Academic.

Jasanoff, S. 1997. Civilization and madness: The great BSE scare of 1996. *Public Understanding of Science*, 6, 221–232.

Kerr, A., Cunningham-Burley, S. & Tutton, R. 2007. Shifting subject positions experts and lay people in public dialogue. *Social Studies of Science*, 37, 385–411.

Kitzinger, J. 1994. The methodology of focus groups: The importance of interaction between research participants. *Sociology of Health & Illness*, 16, 103–121.

Lane, S., Odoni, N., Landstrom, C., Whatmore, S., Ward, N. & Bradley, S. 2011. Doing flood risk differently: An experiment in radical scientific methods. *Transactions of the Institute of British Geographers*, 36, 15–36.

Lang, T. 2014. Chicken contamination: Public should stop buying poultry. *The Guardian*.

Latour, B. 2005a. From realpolitik to dingpolitik: Or how to make things public. *In:* Latour, B. & Weibel, P. (eds) *Making Things Public: Atmospheres of Democracy.* Karlsruhe, Germany/Cambridge, MA: ZKM: Centre for Art and Media Karlsruhe/MIT Press.

Lynteris, C. 2015. Geographies of Contagion, Logics of Containment. Plenary talk at ASA15: Symbiotic anthropologies conference, 14 April 2015.

MacDonald, M. 2013. Ethics of public health. *In:* Storch, J.L., Rodney, P. & Starzomski, R. (eds), *Toward a Moral Horizon: Nursing Ethics for Leadership and Practice.* Toronto: Pearson Education.

Marres, N. 2005. Issues spark a public into being. A key but often forgotten point of the Lippman-Dewey debate. *In:* Latour, B. & Weibel, P. (eds), *Making Things Public: Atmospheres of Democracy.* Karlsruhe, Germany/Cambridge, MA: ZKM: Centre for Art and Media Karlsruhe/MIT Press.

Martin, E. 1995. *Flexible Bodies: The Role of Immunity in American Culture from the Days of Polio to the Age of Aids.* Boston, MA: Beacon Press.

McGoey, L. 2012. Strategic unknowns: Towards a sociology of ignorance. *Economy and Society*, 41, 1–16.

Michael, M. 1996. Ignoring science: Discourses of ignorance in the public understanding of science. *In*: Irwin, A. & Wynne, B. (eds). *Misunderstanding Science? The Public Reconstruction of Science and Technology.* Cambridge: Cambridge University Press.

Nichols, G.L., Richardson, J.F., Sheppard, S.K., Lane, C. & Sarran, C. 2012. *Campylobacter* epidemiology: A descriptive study reviewing 1 million cases in England and Wales between 1989 and 2011. *British Medical Journal Open*, 2. doi:10.1136/bmjopen-2012-001179

Phillips, L., Bridgeman, J. & Ferguson-Smith, M. 2000. *The BSE Inquiry*: vols I–XVI. London: HM Stationery Office.

Rose, G. 1981. Strategy of prevention: Lessons from cardiovascular disease. *British Medical Journal*, 282, 1847–1851.

Royal Society 1985. *The Public Understanding of Science.* London: The Royal Society.

Selinger, E. & Whyte, K. 2012. Nudging cannot solve complex policy problems. *Journal of Risk Regulation*, 1, 26–31.

Shove, E. 2010. Beyond the ABC: Climate change policy and theories of social change. *Environment and Planning A*, 42, 1273–1285.

Simmel, G. 2010 (1903). The metropolis and mental life. *In*: Bridge, G. & Watson, S. (eds), *The Blackwell City Reader, 2nd Edition.* Chichester, UK: Wiley.

Thaler, R. & Sunstein, C. 2008. *Nudge: Improving Decisions about Health, Wealth, and Happiness.* Stanford: Yale University Press.

Thrift, N. 1996. *Spatial Formations.* London: Sage.

Verweij, M. & Dawson, A. 2007. The meaning of 'public' in 'public health.' *In*: Dawson, A. & Verweij, M. (eds), *Ethics, Prevention, and Public Health.* New York: Oxford University Press.

World Health Organisation 2003. *Summary of Probable SARS Cases with Onset of Illness from 1 November 2002–31 July 2003* http://www.who.int/csr/sars/country/table2004_04_21/en/Accessed 4 June 2015, Geneva, Switzerland: World Health Organisation.

World Health Organisation 2009a. *Swine influenza: Statement by WHO Director-General, Dr Margaret Chan,* Director-General of the World Health Organization, 25 April 2009 http://www.who.int/mediacentre/news/statements/2009/h1n1_20090425/en/Accessed 14 September 2015, Geneva, Switzerland: World Health Organisation.

World Health Organisation 2009b. *World Now at the Start of 2009 Influenza Pandemic: Statement by Dr Margaret Chan,* Director-General of the World Health Organization, 11 June 2009. http://www.who.int/mediacentre/news/statements/2009/h1n1_pandemic_phase6_20090611/en/Accessed 14 September 2015, Geneva, Switzerland: World Health Organisation.

Wynne, B. 1992. Misunderstood misunderstanding: Social identities and public uptake of science. *Public Understanding of Science*, 1, 281–304.

Wynne, B. 1995. Public understanding of science. *In*: Jasanoff, S., Markle, G.E., Peterson, J.C. & Pinch, T. (eds), *Handbook of Science and Technology Studies.* Thousand Oaks, CA: Sage.

Chapter Eight
Knowing Birds and Viruses – from Biopolitics to Cosmopolitics

When we shift focus from pathogens to the multifaceted production of pathogenicities, attention becomes a key issue. As we have seen in previous chapters, tending to a disease situation also becomes a multi-faceted occupation.

From the pig farmers and veterinarians who practised immunity through patching together healthy bodies, to the meat inspectors who managed to regulate food through its multiple logics and diagrams, to the people we met in the previous chapter who actively used 'not knowing' amongst other things to cope with the surfeit of information on health; when it comes to pathological lives, attention matters.

Yet, in its hypertrophic form, attention can turn upon itself. The stress of taking care can produce its own ill effects. Likewise, the propensity for immune systems to redefine self as non-self can result in hyper-vigilant or auto-immune diseases. As a state of being, attention can in that sense become symptomatic of too much control. Given these warnings, what kind of attention is appropriate and to be fostered when it comes to pathological lives?

This chapter focuses on the politics of sense and attention. It builds on previous chapters and draws on fieldwork with wildlife experts and virologists to formulate a shift in how we think about attention. In short, this involves a move from a biopolitical understanding of surveillance – where the aim is to police norms and flag up outside threats – towards a cosmopolitical approach to sensing where deviations 'force us to think' and act on the disease situation (Schillmeier, 2013; Stengers, 2010a). These deviations are, we argue, always more than human

Pathological Lives: Disease, Space and Biopolitics, First Edition. Steve Hinchliffe, Nick Bingham, John Allen and Simon Carter.
© 2017 John Wiley & Sons, Ltd. Published 2017 by John Wiley & Sons, Ltd.

affairs, and it is the ability to sense with others that becomes a key element of this cosmopolitics.

We start with an account drawn from Michel Serres (2008) of the more-than-human act of sensing or paying attention. We read Serres as fundamentally opposed to a biopolitics or security apparatus wherein norms are policed. Using wild bird surveillance for highly pathogenic avian influenza as our example, we look at the practices of field and laboratory science. We develop an empirical account of the moment when avian sensitivities were drawn into a 'viral cloud' (Lowe, 2010) of highly pathogenic avian influenza. The gentle science of ornithology took on a strategic urgency once avian migrations changed from being a source of wonderment to a source of potential danger. And, yet, as we note in the later section of this chapter, "Knowing Viruses", knowing birds and knowing viruses were not simply added into the government of things. Instead they opened up a different kind of life politics, where liveliness and sensing were affirmed as eventful in and of themselves rather than as a means to pre-specified biopolitical ends.

Our argument is that these practices share something with a cosmopolitical 'loss of self assurance' (Stengers, 2005a: 996) and it is this loss that needs to be foregrounded as a means to be faithful to the event that is pathological life.

Sensing Life

For Aristotle, and for the philosopher of science Michel Serres, there is nothing in the mind that has not already been sensed. And, for Serres (2008), sensing is everywhere – it is done with many and through others. The nervous system is, it turns out, anything but centred. It is dispersed, distributed, but more than that it defies any straightforward geometrical mapping or system diagramming. Rather, we could say that being sentient is spatially and temporally composite, made up of intensities, events and atmospheres. And, in being so, it is also more than human. It involves or folds together a suite of other bodies.

This creaturely sentience, this redistribution of sensing and knowing, has consequences for a politics of life. The overarching question is, what does a non-central, nervy assemblage, with its wide supportive cast of characters and props, do for a reconfigured, less anthropocentric, more lively, and indeed noisier, politics of life?

The specific issue for this chapter is 'knowing birds and viruses' – a phrasing that, in English at least, is ambiguous and potentially unsettling. At first blush it signifies the activities that are necessary for people to know things (*what* are they, what do they do?). But knowing birds and viruses can also trouble any hard and fast divisions between human knowers and non-human knowns. These knowing entities are just that: birds and viruses *who* know and make collective knowledge possible. Indeed, birds and viruses are understood here to be active players in knowing worlds rather than passive bearers of features available to knowledge. We start this shared sense of knowledge with avian wisdom, before returning to viruses later in the chapter.

Avian Wisdom

The appearance, departures and re-appearance of birds have long been a matter of human delight, foresight and also foreboding. The flight of birds, their seasonal appearances and disappearances, their movements ahead of changes in weather, as well as their airborne vantage, provided for a sense of knowing birds or avian wisdom. The novelist and amateur ornithologist Margaret Atwood summarises the intrigue that bird knowledge, flight and movement has generated for their terrestrial observers:

> For as long as we human beings can remember, we've been looking up. Over our heads went the birds – free as we were not, singing as we tried to. We gave their wings to our deities … and their songs to our angels. We believed the birds knew things we didn't, and this made sense to us, because only they had access to the panoramic picture … a vantage point we came to call 'the bird's eye view'. … Some of us once believed that the birds could carry messages, and that if only we had the skill we'd be able to decipher them. Wasn't the invention of writing inspired, in China, by the flight of cranes? Thoth, the Egyptian god of scribes credited with the invention of hieroglyphic writing, had the head of an ibis. In the ancient world, an entire job category grew up around bird reading: that of augury, performed by seers and prophets who could interpret the winged signs. … 'A bird of the air shall carry the voice,' says Ecclesiastes, 'and those that have wings shall tell the truth' (Atwood, 2010).

Indeed, there was a time when people did nothing without first observing what birds had to say. Soothsayers and Aruspices would, in Roman times for example, consider bird flight or the way they pecked the ground before sanctioning a matter of empire. As Serres (2008: 99) construes it, the greatest empire of all time put its fate in birds. As Machiavelli (1996 (1517): 28–29) noted, this was a key institution of Ancient Rome. For example, *Pollari*, or the guardians of the sacred fowls were required to ask birds for their assent prior to any major enterprise:

> Whence the Romans cared more for this than any other institution, and used it in their Consular Comitii, in starting their enterprises, in sending out their armies, in fighting engagements, and in every important activity of theirs, whether civil or military: and they never would go on an expedition unless they had persuaded the soldiers that the Gods promised them the victory.

Observing or consulting birds was an addiction. Addicere meant to 'speak in favour of' or, in augury, it signals the assent given by birds. For Serres, more radically, this assent is key, and provides the pre-conditions for human speech. Ad-diction, or 'towards-speech', signifies that the ability to speak accrues through the movements of others.

Indeed, the speech act, so triumphantly associated with human rationality and the birth of civilised order, was, for Serres, predicated on birds. As Livy, the Roman historian, stated, nothing could be instigated without the favourable signs or addiction of birds.[1] Rome itself was founded upon the noisy flyby of avian life. The pinnacle of the ancient world, the triumphant empire, was inscribed with non-human nature. Augury was a pre-condition for the city, or the polis. Indeed, augury is the 'feral, non-historical origin of the city' (Serres, 2008: 99), and the polis in this sense was always already a biopolis.

As is characteristic of Serres's philosophy more generally, we should engage augury or the sensing of birds as not so much a quirky historical example of how things used to be (something that Machiavelli's (1996 (1517)) dismissive reading of the *Pollari*, or chicken men, tends to suggest), but as a lesson in how sense and sentience is made. Here, the before-speech noises and movements of birds echo Serres's other works on the non-human foundations of human civilisation (Serres, 1991) and his similarly irreverent histories of science, wherein non-humans and technologies mediate knowledge in ways that de-centre human subjects (Serres and Latour, 1995). So rather than an arcane practice, augury is, for Serres, an exemplary case of a general characteristic of the acts of sensing and knowing. To know and know well, for Serres, involves a pre-engagement with a world of movement, a forecast, or a leap into the world and an entangling of different kinds.

Read in this light, augury and soothsaying need rescuing from disdain. Indeed, these practices, which become symptomatic of superstition and myth in the modern era, are on the contrary exemplary for Serres. In observing the avian world prior to human speech, soothsayers 'are already acting like scientists' (Serres, 2008: 102). In presupposing a world that pre-exists language, where 'meaning manifests itself without us ... [and o]bserving the world as though it were not something brought into being by the collective' (Serres, 2008: 102), augury alerts us to the more than human components of sensing and knowing.

For Serres there is a precious naïvity here with respect to the reality of the world beyond human construction. 'Without being able to prove it I believe, like soothsayers and haruspices, and like scientists, that there exists a world independent of men' (Serres, 2008: 102). And it is this worldliness, this Aristotelian notion that there is nothing in the mind that has not first been sensed (Connor, 2008),[2] that underlines a knowing that is still to be cherished and nurtured.

This sensing of bird sense also returns us to the familiar thread in Serres's writings on communication and the requirement for noise in order for there to be information (Serres, 2007). The noise of the birds is not, for Serres, a signal or message that is to be read or heard without mediation. It is rather the flurry of activity through which, and partly in spite of which, speech is made possible. In brief, the act of hearing or responding to atmospheric waves, vibrations and movements is always more than a matter for the communication of clearly bounded information. Noise is required for sense to be discerned. Communication involves echo and reverberation, a chattering of more than human voices, a continual diffraction of waves. The noisiness of communication, the communing with the world that is necessary for making community possible, is configured in the sensory apparatus of human beings. This is a noisy sentience where the human ear reverberates with the sound waves of others. The ear, that intricate labyrinth of folded skin and nerves, forms a resonance chamber that is the very possibility for language, for theatre, for tribunal:

> It is via this aperture that our eyes turn towards the world, our hearing heeds sounds other than those of language, noises other than those of vocalising. The sounds of scratching or pecking, or the soft caress of feathered wings in the turbulent air – not even Rome could be founded before this movement was heard (Serres, 2008: 100–101).

While Romulus may have visually read the signs of flying birds as assent for the founding of Rome at a particular site, his head was first turned by the caress of feathered wings in the air. Hearing, rather than sight, is the precursor for the biopolis. And sensing movement, through the vibrations that touch the ear, is something that augury preserved and amplified as a moment or event prior to the speech act.

There are at least two issues in all of this. The first is that different bodies sense the world after their own fashion. So, for example, Herbert Stanley Redgrove's twentieth-century account of augury suggested that various species of birds are susceptible to electrical and barometric changes in their atmospheres, 'too slight to be observed by man's unaided senses' (Squier, 2006: 69), and that only this reading of avian sensitivities can explain phenomena like the timing of seasonal and pre-storm migrations. These skeletal and embodied features of avian lives have long since been of interest to those who can gain from accessing the sensitivities of others through various forms of body-snatching. More broadly, it is a practice that is apparent in recent military and industrial investments in bio-mimicry, or the representation and imitation of non-human capacities for human ends (Johnson, 2010).

This extension of human sense through engaging others' sensitivities is of course not what interests Serres. As we have hinted, augury takes us further than the bodily capacities or affordances of other creatures with their varying skeletal and tissue compositions that can undoubtedly make other kinds of sense. So, and second, it is the very matter of making sense as a composite activity that is augury's true promise and Serres's concern. This is more than another account of human sense-making though, albeit one that is enabled or extended by other species or the prosthetics of attachment. As the social psychologist and science studies scholar Steven Brown puts it, there is something going on here that is more than a phenomenological understanding of how sense is made:

> Serres strives for a more radical rethinking of subject and object. It is not simply that we experience ourselves through sensation, it is that what we call 'self', the nexus through which knowledge, feeling and memory are intertwined, is literally there in the midst of things (Brown, 2011: 165).

Brown continues by quoting Serres's (2008: 76) statement that 'the thinking I quivers along the spine, I think everywhere.' 'I think everywhere' is a displacement of the subject, a nervous system that is not so much central, but dispersed. And in being dispersed, there is a folding together rather than an extension of the subject.

There is a geography to knowing in this that is not simply extensive, or about the effectivity or *interessement* or networking, and does not simply act to incorporate others in the process of getting to know or in the act of sensing. The spatiality is not about reach, or amenable to a geometry of the self, rather it is a topological matter, a 'blooming into life' as Brown (2011: 164) captures it. Sensing occurs in the middle of intra-active engagements. Intra-action, as we have noted in earlier chapters, emphasises the generativity of relations and that all parties are altered as relations take shape (Barad, 2007). Knowing birds, then, is not a matter of humans reaching out in order to sense what birds sense, but a matter of knowing *with* a whole suite of sensings. If there is nothing in the mind (knowledge) that has not already been sensed, then knowing birds is a matter of collective sensibilities. To sense at all is already to be in amongst things, to be in the process of intra-acting. Serres is pushing us here to move beyond a common reading of actor-network theory which can, in some hands, focus upon the steady accumulation of heterogeneous matters (Thrift, 2008). For Serres, effective action is a matter of intense moments, repetitive but differentiating encounters where the game is not so much an accumulation and arrangement of bodies, but the ongoing chatter (viral and otherwise, see Chapter 3) and responsiveness to others that makes being sensitive dependent on marking and remarking events. Augury and science, for Serres, are not about extension and effective networks, but about learning to be affected and to be affective in an eventful, differentiating world.

Finally, for Serres, the augury lesson is one that requires relearning. For we have forgotten what it means to pay heed to anything but (human spoken) language. Comparing Roman with contemporary world leaders, Serres satirises it thus:

> I would love to see those who claim to hold the destiny of the world in their hands, whose images we see and whose voices we hear ten times a day, in these times when politics has been reduced to publicising the State – I would love to see them down in the farmyard, their brows furrowed, meditative. Oh to see them thus, standing and gaping in anticipation! The tragedy of speech collapses in laughter (Serres, 2008: 99).

The tragedy of speech is contained in the double meaning of addiction. While to be addicted is to be given assent to speak (through sensing the world and through giving favour), addictions are also a sentence, a dependence. Herein lies the tragedy for Serres, for there is nothing more noisy and disturbing than endless speech-making (those voices we hear ten times a day, publicising the state). So while augury was once the subject of laughter for the rational philosophers, it is now the turn of language and speech to collapse in laughter. Indeed, pushing this a little further, it is the tragedy of a failure to engage with the addictions of avian and other non-human lives that could, in an age of avian influenzas and other zoonotic diseases, prompt some soul-searching of its own. Before we return to this tragedy of too much speech and too little noise, and before we reflect further on this tragic failure, we need to say a little more about how this interest in addiction opens up a livelier and nosier politics of life.

A Livelier Biopolitics and a Noisier Sentience

As we sketched in Chapters 1 and 2, biopolitics was a term coined by Michel Foucault (1981) to mark a shift in emphasis in the government of human societies (Cambell and Sitze, 2013; Lemke, 2011). It signalled a broadening of the techniques and apparatus of government from the disciplining of individual bodies to the knowledgeable manipulation of a population, of statistics and, crudely, the use of precise mechanisms and processes associated with broad-scale changes as a means to affect the direction of those changes (Foucault, 1981: 137). The broad story is that nascent life sciences and statistics, along with techniques honed in plant and livestock management, became intricate with politics in the guise of a range of regularised processes that could be more or less effectively arranged in order to achieve an optimal result. A commonly noted point is that the term and the mode of governing applies largely to the management of society, and to humans, and seeks to render the material or non-human world as predictable, at least in terms of a population or risk pool.

Even so, and somewhat in contrast, biopolitics is also always a more-than-human endeavour. Indeed, as we noted in Chapter 2, populations necessarily incorporate all manner of physical, environmental, biological and economic processes and 'things'. Indeed, the paradigmatic space of biopolitics was not so much the human body, the territory or the nation state, but the circulation of people, goods, finance and so on (Dillon and Lobo-Guerrero, 2008; Foucault, 2007). This feature makes biopolitics rather less about the government of 'men' and more about the government of 'a complex of men and things' (Foucault, 2007: 96). This more relational, less determinate reading of biopolitics prompts both sociologist Thomas Lemke (2015) and the geographer Chris Philo (2012) to suggest that Foucault's work be read in a less structuring, more lively or eventful manner. So we should not necessarily equate biopolitics with power over, or control, or government, or necessarily with a narrowly anthropocentric logic, but with a contingent and always to be worked out 'intrication' of humans and non-humans.

The problem here may not be so much to do with the *means* of governing life, which need, in this version of life politics, to be open to the contingencies and liveliness of humans and non-humans. The difficulties may be wrapped up instead with the *ends* of biopolitics – the optimisation of (human) life through a regulation of circulation. This goal is, for some commentators, rooted in the Heideggerian distinction between proper and improper life (Agamben, 2002; Campbell, 2011; Wolfe, 2013), a distinction that, for Wolfe in particular, results in the separation of 'truly' human life from sub-human and animal life, and sanctions what he calls the non-criminal putting to death of people, domestic and wild animals. In other words, optimising life seems to involve the designation of a norm, a proper life, and once this is settled, all other forms of life are liable to being rendered of lesser worth and, in the case of many of those lives, less lively. Some lives then are disqualified to the point of death in the name of optimising those designated as 'proper'. The familiar if indeterminate dichotomy of human and animal, political life and bare life follows. An ultimate result, for Wolfe at least, is development of the concentrated animal feeding operation (CAFOs – see Chapter 4):

> [T]he practices of maximising control over life and death, of 'making live' in Foucault's words, through eugenics, artificial insemination, selective breeding, pharmaceutical enhancement, inoculation and the like are on display in the modern factory farm as perhaps nowhere else in biopolitical history (Wolfe, 2013: 46).

These foundational and continuously performed distinctions start to unravel of course when we return to Serres's notion of creaturely sensing. Distributed and relational notions of being human, speaking and sensing provide for a relational ethics (Whatmore, 1997), that undermines any straightforward distinction between humans and others. If birds helped to found Rome, and if speech acts are addicted to non-human movement, then the proper and the improper cannot be divided thus. Non-humans are not predictable for humans. Being human is instead predicated on non-humans.

Even so, avoiding this disqualification of many lives in the name of proper life is not straightforward, particularly when biopolitics is often read as a matter of optimising specific (often pre-specified or at least already approximate) forms or ends. It is clear for example that simply stating that people and things are 'intricated' is insufficient for a different kind of biopolitics. Likewise, a blanket affirmation of or pure hospitality to all forms of life is neither practical nor a way of living better pathological lives. We need, in short, a better spatial imagination than one that veers between an exclusive focus on human being or a general dispersal of all being.

As we have stated in Chapters 1 and 5, immunitarian thinking may offer a useful trope in the search for a spatialised understanding of life. Immunity is borne from something other than pure hospitality, and Esposito (2008, 2011) starts to outline a process whereby living requires neither inclusion nor exclusion, but a continuous intra-active learning process. Spatially, this is a mixed affair. Immunity is a matter of communication and marking of difference rather than everything becoming the same.

This non-exclusive exclusion, as Esposito termed it (Chapter 5), is something that Haraway (1993: 306) was keen to evoke in her use of the term 'borderlands':

> The pleasures promised here are not those libertarian masculinist fantasmics of the infinitely regressive practice of boundary violation and the accompanying frisson of brotherhood, but just maybe the pleasure of regeneration in less deadly, chiasmatic borderlands.

Like the immune process wherein poison and cure can amount to the same thing, the trick is to find means of learning by engaging with, and being exposed to difference. There cannot in this sense be a foundational separation between the proper and the improper. Rather there is community that is always already based on the immunitarian. So, writing of the immunological conundrum that is pregnancy in mammals, in a phrasing that would please Serres, Esposito writes:

> The fact that the genetic heterogeneity of the foetus rather than its genetic similarity is what encourages the mother's immune system cannot be reduced to the simple function of rejecting all things foreign. If anything, the immune system must be interpreted as an internal *resonance chamber*, like the diaphragm, through which difference, as such, engages and traverses us (Esposito, 2011: 18, emphasis added).

This recurrence of the resonance chamber, this noisy, amplificatory, intensifying and eventful figure, through which difference traverses us, provides some kind of opening for working together a different kind of life politics with the activity of being or becoming sentient together. So we want to return to the laughable politicians with furrowed brows standing in the farmyard, and to knowing birds, to indicate where this might go.

A Perceptual Ecology of Knowing Birds

As we have seen throughout this book, the last few decades have been punctuated with disease events, which have drawn the state into the farmyard. From mad cow disease to bovine tuberculosis, and severe acute respiratory syndrome to highly pathological avian influenza, the fortunes of government ministers and even prime ministers and presidents have risen and fallen as a result of their ability to handle the labyrinthine complexities of trade, animal health, risk, public health and countryside (Hinchliffe, 2001; Law, 2006). So in this sense politicians *are* on the farm, brows furrowed. They too are concerned with matters of empire, with powers to act. And scientists accompany them, trying to discern a world independent of people. And yet, as Serres laments, perhaps those politicians have forgotten the art of composition, an art that would make those repeated visits to the farm less and less like déjà vu. How, then, to sense differently and to develop a different kind of life politics? Let us return to the birds, to the wonders and threats of avian and viral circulation as it developed once the viral cloud called H5N1 started to move over the UK.

In autumn, on the wetlands of the British Isles, bird people eagerly await the return of seasonal migrating birds. Lying on the edge of the Eurasian land mass and warmed by the Gulf Stream, these islands make a particularly favourable destination for winter migratory visitors from Northern Europe, Siberia, Canada and Iceland. Indeed, no equivalent-sized area on Earth draws its wintering birds from such a wide latitudinal span (Newton, 2010). Come spring, wildfowl and waders relying on winter lakes that seldom freeze over will make their way to higher latitudes where the lengthier days make for rich feeding and nesting grounds. At that time of year, other birds, from as far away as South Africa and the Antarctic, will start to migrate north to Great Britain and Ireland, taking advantage of the mild summer and the new abundance of food.

It is easy as a researcher amongst ornithologists to sense the anticipation as the winter migration approaches. News passes from birders in the near continent, in the Netherlands and Belgium: birds are on their way. Bird people watch the skies for known birds and are visibly excited and relieved when they arrive 'home'. The differences from year to year in migratory arrival dates, with early birds returning in October and late ones inching into November, provides a side show of endless discussion about normal annual variation and weather patterns, and the more sinister possibility that there are noisy signals of a shift in climate. Arriving birds may be the harbingers of bad tidings with respect to the coming winter or even shifts in climate, but the east winds that can accompany their arrival are the only similarity to Du Maurier's famous short story 'The Birds'. Any foreboding in the offices of the Wildfowl and Wetland Trust relates to concerns over the plight of individual birds, known to reserve workers for years or even decades. Their lives and deaths are related to conservation concerns, to reductions in wetland habitat along the 'corridors' that migrating birds use and which are vital stopovers and feeding stations on their long and arduous journeys. There is a curiosity to see which birds have returned, how they have fared and what social relations they have maintained or started. The sense is one of hospitality to the wild migrant, and an oft-expressed wonder at the capacity of birds to move long distances, to return to the same places and to maintain social groupings through intercontinental migrations, sometimes despite the odds.

This hospitality is underwritten with a suite of activities that contribute to the knowing of migrations. This knowing is, as Serres might have anticipated, a composite activity that enacts a sensuous world, a world that 'blooms into life'

(Brown, 2011: 164). In other words, knowing birds is not a matter of human subjects getting to know their avian objects through a representational economy so
much as a process of human-avian intra-actions. Sensing birds involves a suite of
embodied activities, bodies arched forward, lenses at hand and to eye. There are
rapid movements of eye muscle, retinas primed to 'jizz', a mode of recognition
based on the slightest of 'signals', be that a silhouette, a fleeting glimpse, a reading
of proportion or a characteristic movement – it is a semiotics of the fleetingly
readable. In short, there is an embodied readiness to the world, an affective
attunement to avian bodies and movements (Despret, 2004; Law and Lynch,
1990; Lorimer, 2008). Using these learnt if not easily codified ways of knowing,
species are identified usually without a thought. The material semiotics of the
fleetingly readable are added to and made sense within a set of other avian
knowledge practices. There are counts, observations, ringing and tracking, and
we will say something about each in turn, but will focus for the main part on
observation.

The following field diary entry evokes the activity that takes place as part of a
routine day on a wetland reserve in Gloucestershire, England:

> Up in an observation tower, a reserve warden counts Siberian white fronted geese
> grazing in the fields along the estuary. 'They're a flagship bird. Slimbridge has one of
> the biggest flocks in the UK, so we try to count them daily on our rounds.' He's also
> quick to spot new and unusual arrivals. 'I know this place like the back of my hand.'

So birds are identified, sometimes counted in order to compile population
data, and the out of the ordinary are quickly registered. The counts serve all
kinds of knowledge-building programmes, furnishing longitudinal studies of
species numbers. In addition, some species of bird are observed in detail, a
practice that allows for identification of individuals and social groups and
helps to generate records of a bird's life, habits and relations. It is a practice
that may have a long and unwritten history, but it is notable that here charismatic species become entangled with charismatic and foundational figures in
the twentieth-century conservation movement. These 'visionary' pioneers
managed to learn to be affected by morphological differences and generate a
mode of recognition based less on the fleeting glimpse and more on studied
observation. So, Sir Peter Scott, the huntsman turned conservationist, artist
and founder of the Wildfowl and Wetlands Trust (WWT), learnt through the
careful observation and painting of birds that each Bewick's swan had an
individual bill patterning. From the late 1950s, having successfully attracted
migrating Bewick's swans to Slimbridge on the River Severn in Gloucestershire,
he started to sketch every Bewick visitor to the reserve (see Figure 8.1). Noting
in his diary that this minute and detailed observation 'may or may not be
scientifically useful',[3] this is now one of the longest running single species
studies in the world.

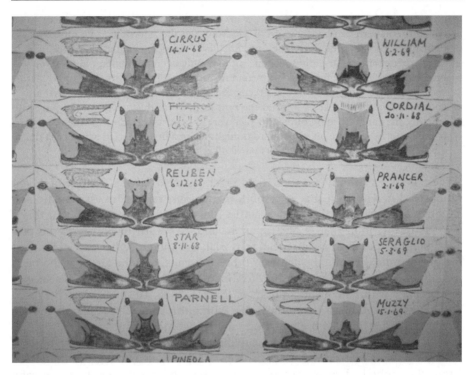

Figure 8.1   Scott's swan sensorium. (Photo by Hinchliffe – Image used with permission from the WWT and Scott Foundation.)

Scott's sketches may appear at first sight to be nothing more than an identification chart, a series of mug shots. But this is to miss the shift its production configures in the ways in which bird lives are sensed. No longer a mass of undifferentiated life, to be hunted or otherwise, the procession of swans contributes to a reshaping of the ways in which humans start to engage with avian lives. It is, then, a sensorium of avian life – it and its production contribute to a shift in the sensing of life.

Scott noted each swan's arrival and/or return in autumn, as well as each bird's habits and social behaviour. The birds were differentiated on the basis of their beak patterns, but at the same their associations and relations became matters for study. And these detailed observations are still practised over half a century on, with certain staff maintaining the painstaking work of learning to sense individual birds and record their movements and social relations (their pairings, partings and battles), and adding to this database any observations of health, status and changes in demeanour. The following extract of field notes gives a sense of this activity:

> From a pile on her desk, she pulls a clipped bundle of A5 papers. Each sheet represents an individual bird, and has been produced from a template. There are fields for

notes about the individual bird, and blank outlines of a swan head and beak on which the pattern can be drawn from the front and the side view. 'The process of drawing helps you remember,' she says. She also prefers drawing as it's faster than setting up a camera for a photo when you may only get a momentary glimpse of a bird. ... Every year she tries to identify and record all wild swans over a year old that visit the reserve. Birds that she can't recognise fall into two categories: those that are confident, and those that fly away at the bird feed. From this behaviour she judges whether they are returning birds, in which case she must work through the records of bill patterns to figure out which bird it might be, or a first-timer, in which case she gives the bird a name and records its particulars.

What Scott's sensorium and these field practices make clear is a mode of recognition or differentiation that is careful, meticulous, repetitious, systematic and storable. Moreover, for the social scientist in their midst, it is notable that this is a mode of attention that is quiet, patient and, to return to Serres, is the precondition for speech. It is, to repeat, sensorial, a folding together and a blossoming of avian life. Despite the artifice of the landscape, and the fact that the birds would not be here at all if it was not for Scott's initial work, through this science or augury, 'a world independent of men' (Serres, 2008: 102) flickers to life.

The point to emphasise for now is that within this suite of practices there is an affective register that is more than body snatching, more than simple representation. There is a knowing together, a folding together of twitchers and birds, a becoming avian even, which is born from a repetitious making of differences. From the embodied memory work of sketching beak patterns, to the careful attachment of devices and rings to delicate feathered bodies, there is a sensorial link made between avian and human bodies. The links are perhaps not quite ones that entail a move from avian lives to Roman gods, and so fearful worship. But there is a linking together of experience and cognition that is productive of a reverence for avian life. Ornithology is then more than a matter of accounting or drawing avian lives into the world of people. It is a perceptual ecology that involves intra-actions of people and birds.

Surveying Life

If the observational science of ornithology is characterised by hospitality, the welcome returns of travellers signalling another year of a frail order, then the advent of highly pathogenic avian influenza threatened to overturn these established modes of recognition, with hospitality giving way to hostility. A now familiar if over-stated story is the resulting transformation of wetlands into frontlines in the fight against the incursion of disease, with ornithologists becoming part of an ever extending border agency in the fight against pandemic influenza (Braun, 2007; Wilbert, 2006). In biopolitical terms, the arts of ornithological composition have seemingly been extended to the government of things, in the name of protecting the extraction of surplus from other avian lives and the safe-guarding of human health. And yet, life and knowing birds is not so easily co-opted. As we now go on to show, the object-rich observational sciences with their more-than-human sentience seem to desist from any easy affiliation to a power over life.

To be sure, with the advent of the viral cloud, migrating birds shifted from being a delight to being, in some quarters at least, a possible danger, an invading presence, even a source of terror.[4] Wild birds traversed and linked spaces and places across Eurasia where highly pathogenic avian influenza was considered to be circulating within wild and domestic bird populations. The migratory maps that ornithologists had produced suddenly represented pathways from avian to human lives through the amplificatory effects of domestic bird flocks.

More generally, hospitality may well be permanently marked by the potential for things going wrong, or for visitations to turn hostile (Derrida, 2002). Indeed, the desire that is embodied in the observations of migrating birds, the alterity, the other worldliness of avian lives, may well be contingent upon the possibility for things to be otherwise. Migrating wild birds may well be both intoxicating and potentially toxifying. This is the unconditional condition for hospitality within Derrida's thought (see also Hagglund, 2008). Hospitality can never be unconditional in the idealised sense as an opening to alterity, but is always haunted by the possibility for things to go awry. That is, the very exposure of human and avian lives to one another is also an exposure to incalculable or indeterminate others. The latter, wildness so to speak, is both the source of threat but also, it needs to be emphasised, a pre-condition for good. So things might go wrong in the hospitable relation, but also, for Derrida, 'nothing would be desirable without' the possibility that the good 'bears the cause of its own destruction within itself' (Hagglund, 2010: 297). The *immunitarian trope* returns here as auto-immunity, but for now it is important to briefly outline the practices that are involved in this making hospitality conditional.

The possibility that the arrival of migrating wild birds might augur an epizootic in domestic livestock and, even worse, transform into a zoonotic disease capable of mutating to become a human pandemic, initiated a set of responses from state and industry that largely focused on biosecurity. In brief, these involved a risk-based analysis of migratory sites and species, a sampling of found dead, shot and healthy wild birds, screening of those samples for influenza, measures to separate wild and domestic species, and ornithological studies of affected sites in order to develop appropriate exclusion and containment policies (Hinchliffe and Lavau, 2013). In effect, the overall programme was one of separating domestic life from wildlife and a *policing* of both in order to provide early warnings of infringements and implement lockdown emergency measures (further containment and suspensions of movements) once the sequestered safe life had been or was in danger of being breached.

In generating risk knowledge through to populating studies with samples and expertise judgements, ornithologists were involved in offering practical expertise and the labour necessary to developing what could be called, in the round, avian surveillance. So for example, in developing an analysis of at-risk sites and species, ornithological experts contributed their knowledge of migration routes and usual feeding and nesting locations. Through longstanding pathological studies, they could estimate which species were more likely to host the avian influenza virus, with or without clinical symptoms. This knowledge, along with coarse-grained information on GB poultry, resulted in a prioritising of certain locations and species as most at risk of introducing HPAI to the UK and then passing it to domestic livestock. Routine ringing practices were then augmented with disease surveillance – the sampling of birds and the taking of biological samples within which the presence or absence of avian influenza was to be determined:

> I join a vet at the swab station. (…) She inserts a barcoded swab, a bit like a cotton bud stick, into the bird's mouth and about 6 cm down the throat. A gentle twist and then out, into a plastic cylinder with a stopper top. This is placed in a box with the others – the same is then done for the anus. … [T]he birds are then weighed, and we then carry them through a gate, behind a hide, and release them back on to the mere.
>
> The swab samples are carefully placed into boxes, onto which are attached symbols indicating hazardous materials, and they will go back to Slimbridge before going off to Weybridge in Surrey for analysis. (Field diary entry)

The bird survey was a repetitive and costly process. In order to generate a reasonably reliable signal of avian viruses in wild birds, thousands of birds needed to be caught every year. But the work did not stop there. Knowing viruses needed to be added to knowing birds.

Knowing Viruses

As we noted in Chapter 3, influenza viruses are highly mutable and exist in heterogeneous clusters or swarms. They are characterised by their differences so that any representation of *a* virus is always an approximation of a population of viruses or quasi-species. Thus a 'consensus sequence' of a particular viral strain may, depending on sequencing technology and statistical approaches, involve large error bars with respect to the variance that exists in any viral swarm. This variance may well play a key role in the development of new strains as well as medicine and vaccine resistant viral variants (Chapter 3). In other words, knowing viruses is more than a matter of identification. It involves understanding the indeterminacy of that knowing.

Added to this difficulty in knowing what a virus is, there is also the difficulty in knowing exactly what a virus can do (see Chapter 3 on this declaration of ignorance). For knowing viruses involves knowing their situation, understanding how the quasi-species is pushed and pulled by other bodies and environments.

Surveying for viruses is troubled in this sense by the twists and turns of a topological existence.

Despite this need for a knowing around of viruses, there are tensions between the need to detect viruses and the requirement to develop viral knowing. Screening the materials gathered in the field tends to involve the attempt to recognise forms that have a past history of infection (Hinchliffe, 2015; Hinchliffe and Lavau, 2013). This is, in short, a search for genetic fingerprints, either in the form of viral material in samples from swabbed excretory materials or through antibodies present in blood samples, and is dependent on a particular epistemology – one that is implicitly focused on Koch's postulates (Chapter 2) or a single cause for a single disease model (for a history, see Attenborough, 2011).

The process of extracting virus materials from field materials is indicative of this pathogen-focused approach to disease. The purification of field materials through laboratory-based cleaning, killing, buffering, spinning, splitting and filtering produces dead strands of viral nucleic acid. These strands are then amplified and 'read' to produce a consensus sequence – with the presence or absence of key genetic segments of the virus indicative of the disease status of wild bird flocks.

Similarly, serology or haemagglutinin inhibition (HI) tests of blood samples seek to trace previous infections, antibodies within the blood of sampled birds that are produced in response to pre-determined antigen (HA or haemagglutinin) proteins on the coats of viral particles. Again the trace of the presence or absence of that pathogen, read through its immunological proxy, the antibody, is the key to surveillance. But, as the scientists involved are all too ready to discuss, these tests are far from perfect.

In terms of serology, responses and antibody production may be subdued or, conversely, 'noisy' as a result of maternally conferred immunity, co-infections or in the case of farmed birds, any previous vaccinations. It is also common for serological tests to produce positive results but for no live virus to be detected in the sample (so-called cryptic infections), suggesting either that infections have passed through or that the antibodies are being produced in response to a complex of conditions and microbial challenges. Meanwhile, the tests themselves need to be continually balanced as they may be over-sensitive, producing false positives, or conversely not sensitive enough, being overly specific to a suite of viral strains and therefore likely to miss any dangerous shifts or drifts in the virus and in immune responses.

For the surveillance of excretory materials, the scientific practice is also more interesting and complicated than the word survey might suggest. If initial screening suggests a possible presence of an influenza A virus (suggested in this case through the presence of the relatively stable matrix gene that codes for the protein coat of the viruses), then the ensuing practice starts to look less like the dusting for fingerprints and more like a detective story wherein the various components of the story are pieced together or coordinated.

Indeed, the samples gathered in the field start to circulate in laboratories where classical rather than molecular virology using *in vitro* and *in vivo* methods and conventional virus isolation, culture and incubation take place. Field samples are inoculated into embryonated fowls' eggs, grown and then extracted, for further testing and characterisation. The field sample thus proliferates, such that it becomes the subject of multiple simultaneous investigations. In this case, diagnosis is not simply a process of elimination; it is a suite of practices that produces new material forms that may gradually conform sufficiently to deliver a probable understanding of what indeed a virus can do (on the simultaneous nature of multiple forms, see Mol, 2002).

The complexity of this disease ecology and the attendant knowledge practices makes surveillance less a matter of detecting known enemies and more a procedure of piecing together evidence from a variety of experiments, learning to respond with various epistemic objects including proteins, viruses, red blood cells as well as population data and disease models, in order to build up a picture of viral circulations. Knowing viruses, in other words, requires know-how as much as a knowing what. And knowing diseases becomes a trans-disciplinary practice rather than a semi-automated process of pathogen identification. This non-specific understanding of disease emerges through the multiple reading and corroboration of often subtle signs within a variety of spaces and settings. It involves piecing together several practices and practitioners, each with different obligations, traditions and embedded uncertainties. Health becomes more than locating disease threats or contaminants, it is understanding the ecologies and configurations of viruses, hosts, responsive bodies, reagents and the stresses of production systems and habitats that matter. In other words, disease surveillance and knowing viruses requires a supplement to the business of a survey. The distinction, often institutionalised, between the science of viral surveillance and virological research become less than clear. Observation, a key part of any knowing, becomes vital.

The Significance of Observation

In these accounts of both field and laboratory science, disease cannot be reduced to simple objects that need to be regulated and monitored. Diagnosis is in this sense a 'knowing around' of a matter.

For Serres, this multi-sensual knowing around is associated with observation as opposed to surveillance, though as we have seen the latter is never quite what it seems – indeed the danger is not one being replaced by the other but the failure to recognise the importance of a knowing around in any claim to know. A world of difference is produced through a sensorial process, a repetitious intensity and a care for objects. While the survey in its baldest sense is a technique that at least in theory is set up to record the presence or absence of notifiable pathogens within wild birds, repetition here does not yield difference, or a new sensorium, but provides confirmation or otherwise of the disease status of bird life. On the surface at least it is a yes/no technique with little room for qualifiers or grey areas. In that sense, there are known identities and categories (viruses and presences), which are being surveyed for transgressions (are they here?), with any positive result invoking a statutory declaration of a notifiable disease with all the legal and international trade requirements for spatial containment.

So, if the observation of birds provides an exemplar of the pre-condition for knowing and sensing, then the survey might well be closer to what Serres (2008: 39) somewhat polemically critiques as a 'surveillance of relationships'. Here Serres attempts to tease apart the observational sciences from what Foucault tended to lump together as part and parcel of a biopolitics of discipline and security, of 'surveillance, observation and correction' (Foucault, 1977: 198). And a key difference for Serres is the status of the object or 'things'. In surveillance we are in the theatre of politics wherein objects 'just are' and the job of the surveyor is merely to police the relations between these known entities. It is a world without time, a Euclidean space of billiard-ball entities (Chapter 3), whose relations are only ones of geometrical distance and proximity. As the geographer Doreen Massey (2005) would say, such a timeless world is also a world without spatial multiplicity, a world where nothing new can happen. It is also a world that moves from (fixed) objects to their relationships, and in the process the object effectively disappears as a matter of interest. In this sense politics is, for Serres (2008: 41), 'object poor'. And when the philosophies without object encounter an object, 'they change it, by sleight of hand, into a relationship, language or representation' (Serres, 2008: 41).

Observation and the sciences, on the other hand, come after the object and 'foster its emergence' (Serres, 2008: 41). Observation is object rich, and is concerned with things. Things, indeed, are in time, historical and, we would argue, geographical. That is they are always in process, partially related – certainly not determined by those relations, least of all by their relationships to people – and spatially multiple. This is one of the key lessons of the last few decades of science studies. To observe and to observe well is to learn to be affected by the world, to be attuned to it (Despret, 2004; Latour, 2004), to engage in building a sensorium and, in doing so, to make differences. Rather than move from fixed and time- and space-less objects to the surveillance and policing of their relationships, observation focuses on things-in-relation, to their configurations and differences. To say it quickly, instead of marking presence and absence of a virus, a noisier sentience attends to the viral cloud. As Lowe's (2010) ethnography of bird flu suggests, and as Wallace's (2009) unique efforts to marry viral dynamics to global capital suggest, this cloud can be thought of as a sensorium of differentiating and swarming viruses, of selection pressures, of avian lives lived at the edge of biological margins.

For Serres (2008: 39), this 'immense difference between the observation of things and the surveillance of relationships', which mark out two worlds and two times in opposition – 'that of myth and that of our history' – is not simply a matter of contrasting modes of address, but also marks out a formative and generative condition for politics and culture. Moreover, Serres continues, the society where surveillance dominates 'ages quickly, becoming old-fashioned and abusively archaic. The past lurks there like a monster, harking back to the age of myth.' But what is it to 'age quickly'? Serres is ostensibly writing about the exclusionary suspicions that lie within a marriage or a community, and the anxiety that generates nothing but premature ageing. But to age quickly is also to become out of date, to assume that any knowledge accrued is unconditional and unchanging. It is to become too focused on the prematurely fixed objects, in this case to 'fix' a pathogen rather than see pathogenicity as a situation, a constantly varying swarm of differentiated elements and relations. It is to seek to police relationships rather than allow observation to engage with the world or intra-actions of things and relations. The broader point is that surveillance requires observation, and this changes the politics from one of regularisation and arrangement of things, to an engagement with the resonance chamber of learning to sense and sense well. In Stengers's terms it is a matter of learning to 'think' rather than to recognise (Stengers, 2005b: 185).

The final point is that the state of anxiety associated with constant vigilance and security (Cooper, 2008; Lentzos and Rose, 2009), that for Serres generates nothing more than ageing, may be rather different to the politics of life that follows from his 'observation of things'. We return to this in the Conclusions, but for now this shift draws us close to the cosmopolitics that we introduced at the start of the chapter as well as in Chapter 1. It is the ability to observe and to find new modes of normativity from discrepancies or differences that marks out this politics. These are what Serres refers to as 'new orders' associated with the noise or deviations that is the parasite (Serres, 2007). Brown (2013: 91) translates these as 'counternorms' or the ways in which 'illness and pathology push the organism to adapt by shifting the normative basis of its functioning'. This invocation of Canguilhem's (1991 (1966)) troubling of any divide between the normal and the pathological returns us to a cosmopolitics of illness (Schillmeier, 2013) and to Stengers's minoritarian politics (Stengers, 2010b). As we expand in the Afterword, cosmopolitics involves the reconfiguring of pathological lives as matters that are not measured by a deviation from the norm (or proper life), but as deviations that may open up the possibility for empowering a disease situation. The addictions of birds, the assent of viruses, is vital to this politics.

Conclusions

We started this chapter with Serres and the soothsayers and auspices of ancient Rome. We ended with a sense of what it might mean to talk again of knowing birds and viruses. To see our politicians in the farmyard with furrowed brows observing birds would be rather different to the fast-twitch responses that characterise the will to draw life into the governance of things. If the soothsayers underlined that the polis has always been a biopolis, then Serres's excursion into sensing reminds us of the addictions that make speech and politics possible. This is a livelier, noisier politics. It is one that challenges not just the means of government but its ends too. In this sense, we might not best view biopolitics as necessarily involving the insertion or intrication of humans, non-humans and otherwise, into an already constituted bio-power. Such powers over life inevitably fail; for Serres, they age quickly and become obsolete. There is something that escapes in knowing birds and viruses, that cannot be grasped, and that renders a policing of relations problematic. Knowing birds and viruses involves a different kind of biopolis, one that continuously reminds us of the more than human conditions for speech. It provides an opening, however hesitant, onto the powers of life and of living together.

Endnotes

1   Livy, 6.41: *auspiciis hanc urbem conditam esse, auspiciis bello ac pace domi militiaeque omnia geri, quis est qui ignoret?* ('Who does not know that this city was founded only after taking the auspices, that everything in war and in peace, at home and abroad, was done only after taking the auspices?').

2   We are relying here on Connor's (2008) excellent and informative commentaries on Serres's collective body of work.

3   The diaries from the earliest days of the Slimbridge project are kept in staff offices at the WWT headquarters in Slimbridge. This is one of the earliest entries by Scott.

4   The distinction between a biological terrorism as orchestrated by enemies of the state and that generated through emerging infectious diseases was, for some, becoming less pertinent in a world that was being more anxious of, if not really prepared for, unpredictable events (see Cooper, 2006).

References

Agamben, G. 2002. *The Open: Man and Animal*. Stanford, CA: Stanford University Press.

Attenborough, F. 2011. The monad and the nomad: Medical microbiology and the politics and possibilities of the mobile microbe. *Cultural Geographies*, 18, 91–114.

Atwood, M. 2010. Act now to save our birds. *The Guardian*, 9 January 2010.

Barad, K. 2007. *Meeting the Universe Halfway: Quantum Physics and the Entanglement of Matter and Meaning*. Durham NC: Duke University Press.

Braun, B. 2007. Biopolitics and the molecularization of life. *Cultural Geographies*, 14, 6–28.

Brown, S.D. 2011. A topology of the sensible: Michel Serres's 'The Five Senses'. *New Formations*, 72, 162–170.

Brown, S.D. 2013. In praise of the parasite: The dark organizational theory of Michel Serres. *Informática Na Educação: teoria & prática*, 16, 83–100.

Cambell, T. & Sitze, A. (eds) 2013. *Biopolitics: A Reader*. Durham, NC: Duke University Press.

Campbell, T.C. 2011. *Improper Life: Technology and Biopolitics from Heidegger to Agamben*. Minneapolis: University of Minnesota Press.

Canguilhem, G. 1991 (1966). *The Normal and the Pathological*. New York: Zone Books.

Connor, S. 2008. Introduction, in Serres, M. *The Five Senses*. London: Continuum.

Cooper, M. 2006. Pre-empting emergence: The biological turn in the war on terror. *Theory, Culture and Society*, 23, 113–135.

Cooper, M. 2008. *Life as Surplus: Biotechnology and Capitalism in the Neoliberal Order*. Seattle: University of Washington Press.

Derrida, A, J. 2002. Hospitality. *In*: Anidjar, G. (ed.), *Acts of Religion*. London: Routledge.

Despret, V. 2004. The body we care for: Figures of anthropo-zoo-genesis. *Body & Society*, 10, 111–134.

Dillon, M. & Lobo-Guerrero, L. 2008. Biopolitics of security in the 21st century. *Review of International Studies*, 34, 265–292.

Esposito, R. 2008. *Bios: Biopolitics and Philosophy*. Minneapolis: University of Minnesota Press.

Esposito, R. 2011. *Immunitas: The Protection and Negation of Life*. Cambridge: Polity Press.

Foucault, M. 1977. *Discipline and Punish: The Birth of the Prison*. Harmondsworth, UK: Penguin.

Foucault, M. 1981. *The History of Sexuality, vol. 1: An Introduction*. Harmondsworth, UK: Penguin.

Foucault, M. 2007. *Security, Territory, Population: Lectures at the College de France 1977–78*. London: Palgrave Macmillan.

Hagglund, M. 2008. *Radical atheism: Derrida and the Time of Life*. Palo Alto, CA: Stanford University Press.

Hagglund, M. 2010. A non-ethical opening of ethics: A response to Derek Attridge. *Derrida Today*, 3(2), 295–305.

Haraway, D. 1993. The promises of monsters: A regenerative politics for inappropriate/d others. *In*: Grossberg, L., Nelson, C. & Treichler, P. (eds), *Cultural Studies*. London: Routledge.

Hinchliffe, S. 2001. Indeterminacy in-decisions – Science, policy and politics in the BSE crisis. *Transactions of the Institute of British Geographers*, 26, 182–204.

Hinchliffe, S. 2015. More than one world, more than one health: Reconfiguring interspecies health. *Social Science and Medicine*, 129, 28–35.

Hinchliffe, S. & Lavau, S. 2013. Differentiated circuits: The ecologies of knowing and securing life *Environment and Planning D: Society and Space*, 31, 259–274.

Johnson, E.R. 2010. Re-inventing biological life, re-inventing the human. *Ephemera: Theory and Politics in Organization*, 10, 177–193.

Latour, B. 2004. How to talk about the body? The normative dimension of science studies. *Body & Society*, 10, 205–229.

Law, J. 2006. Disaster in agriculture: Or foot and mouth mobilities. *Environment and Planning A*, 38, 227–239.

Law, J. & Lynch, M. 1990. Lists, field guides, and the descriptive organization of seeing: Birdwatching as an exemplary observational activity. *In*: Lynch, M. & Woolgar, S. (eds), *Representation in Scientific Practice*. Cambridge, MA: MIT Press.

Lemke, T. 2011. *Biopolitics: An Advanced Introduction*. New York: NYU Press.

Lemke, T. 2015. New materialisms: Foucault and the 'Government of Things'. *Theory, Culture and Society*, 32, 3–25.

Lentzos, F. & Rose, N. 2009. Governing insecurity: Contingency planning, protection, resilience. *Economy and Society*, 38, 230–254.

Lorimer, J. 2008. Counting corncrakes: The affective science of the UK corncrake census. *Social Studies of Science*, 38, 377–405.

Lowe, C. 2010. Viral clouds: Becoming H5N1 in Indonesia. *Cultural Antrhopology*, 25, 625–649.

Machiavelli, N. 1996 (1517). *Discourses on Livy*. Chicago, IL: University of Chicago Press.

Massey, D. 2005. *For Space*. London: Routledge.

Mol, A. 2002. *The Body Multiple: Ontology in Medical Practice*. Durham, NC: Duke University Press.

Newton, I. 2010. *Bird Migration*. London: Collins.

Philo, C. 2012. A 'new Foucault' with lively implications – Or the 'crawfish advances sideways'. *Transactions of the Institute of British Geographers*, 37, 496–514.

Schillmeier, M. 2013. *Eventful Bodies: The Cosmopolitics of Illness*. Farnham, UK: Ashgate.

Serres, M. 1991. *Rome: the Book of Foundations*. Stanford, MA: Stanford University Press.

Serres, M. 2007. *The Parasite*. Minneapolis: University of Minnesota Press.

Serres, M. 2008. *The Five Senses: A Philosophy of Mingled Bodies*. London: Continuum.

Serres, M. & Latour, B. 1995. *Conversations on Science, Culture and Time*. Ann Arbour, MI: University of Michigan Press.

Squier, S.M. 2006. Chicken auguries. *Configurations*, 14, 69–86.

Stengers, I. 2005a. The cosmopolitical proposal. *In*: Latour, B. & Weibel, P. (eds), *Making Things Public. Atmospheres of Democracy*. Cambridge, MA: MIT Press.

Stengers, I. 2005b. Introductory notes an ecology of practices. *Cultural Studies Review*, 11, 183–196.

Stengers, I. 2010a. *Cosmopolitics 1*. Minneapolis: University of Minnesota Press.

Stengers, I. 2010b. Including non-humans in political theory: Opening Pandora's Box? *In*: Braun, B. & Whatmore, S. (eds), *Political Matter: Technoscience, Democracy, and Public Life*. Minneapolis: University of Minnestoa Press.

Thrift, N. 2008. *Non-representational Theory: Space/Politics/Affect*. London: Routledge.

Wallace, R.G. 2009. Breeding influenza: The political virology of offshore farming. *Antipode*, 41, 916–951.

Whatmore, S. 1997. Dissecting the autonomous self: Hybrid cartographies for a relational ethics. *Environment and Planning D: Society and Space*, 15, 37–53.

Wilbert, C. 2006. Profit, plague and poultry: The intra-active worlds of highly pathogenic avian flu. *Radical Philosophy*, 139, 2–8.

Wolfe, C. 2013. *Before the Law: Humans and other Animals in a Biopolitical Frame*. Chicago and London: The University of Chicago Press.

Chapter Nine
Conclusions – Living Pathological Lives

At the start of this book we marked four moves that characterised our interest in pathological lives and the spatial politics of disease. They were, first, a shift to socio-technical diseases; second, a move from disease sites to situations; third, a focus on pathogenicity; and fourth, a reconfigured politics of life. In subsequent chapters we detailed how infectious disease can be understood as an assemblage of socio-technical matters (move 1); how concentrated animal feeding operations, food processing sites, disease publics, changes to food surveillance and so on provide the conditions of possibility for a disease situation (move 2); and how the epidemiological gaze might shift from a focus on pathogens to a concern with the social, economic as well as biological processes that contribute to pathogenicity (move 3).

With this shift in attention, *from* pathogens-and-sites *to* pathogenicities-and-situations, we have also made certain conceptual shifts in understanding and a reconfigured politics of life (move 4). We have employed a style of spatial thinking as a means to better comprehend how pathological lives are made from the interrelations of disease diagrams, materials and practices. And we have identified openings for a different kind of life politics. Attempts to shore up current norms and versions of life seem to generate inevitable failures and spill overs. In their place, we have argued that those practitioners who work with, rather than deny, the material heterogeneity and complex spatiality of pathological lives offer the best resources for remaking safe life and generating a livelier politics of life.

Pathological Lives: Disease, Space and Biopolitics, First Edition. Steve Hinchliffe, Nick Bingham, John Allen and Simon Carter.

In the remaining pages we will draw together our arguments in order to amplify what we mean, in particular, by this spatialised reconfiguration of a lively politics. We end by stating the ways in which this re-orientation changes our approach to the infectious disease emergency that set the scene for the opening chapter of this book.

We start this closing statement by reading our own argument against the grain in order to pose a challenge. 'Pathological Lives' *can* seem to simply replace a uni-causal explanation of infectious disease with an account that simply makes everything, or the system, responsible for disease. Instead of being an acute 'outside' event that affects an otherwise healthy body, infectious diseases or their conditions of possibility are, in this argument, already present within the circulating body. They lie in wait in the poultry houses, flare up in under-pressure pig farms, or take advantage of under-resourced regulatory systems and so on. There is, in this account, not so much a *break-in* of disease, but a *break-out* from a system that is already in that sense pathological. Indeed, here we can return to our use of Foucault's (1973: 153) understanding of pathological life, wherein 'the idea of a disease attacking life was replaced by the much denser notion of pathological life.' This *denser notion* implies a mesh or interpenetration of matters that can produce disease.

We have been keen throughout the book to tease out the ways in which pathogenicity is produced through the interrelation of various social, economic, governmental as well as biological matters. Disease situations arise, we have argued, from these matters and the interplay of disease diagrams. At first blush, this is not far removed from other processual approaches to disease. For example, Singer's (2009: 14) 'syndemic' approach to health 'involves a set of enmeshed and mutually enhancing health problems that, working together in a context of noxious social and physical conditions, can significantly affect the overall disease burden and health status of a population.'

While this syndemic approach shares a good deal with our approach, there is a danger that the inter-meshing of multiple threads or causes produces a view of disease that is itself rather dense, difficult to mobilise and in the end imprecise. Worse, there are elements of this view that could be used to sanction even greater and possibly misplaced investment in social and biological control, some would call it the 'new normal' of a post-millennial governing through fear and insecurity. The risk here is that as *everything* becomes pathological, so every area of life becomes subject to continual and exhausting vigilance. Foucault's 'denser notion of pathological life' can in this sense prove both impenetrable and subject to greater and greater expansion of security.

We have therefore endeavoured to supplement Foucault's 'denser notion' with a more precise lexicon. We have done so by focusing on the *tensions* that exist within and between disease diagrams and the multiplicity of demands that are made on chickens, farmers, inspectors, people, food and so on. Whether or not these demands amplify or reduce pathogenicity has been our key aim. Part of the answer here has been to develop a form of spatial thinking that allows us to understand these tensions, their interrelations, the intensities that these can generate and their ability to provide new conditions for disease. In addition, we have been keen to highlight the hard work of *attending* to and *patching* together healthy lives in these multivalent conditions.

To be sure, part of the issue with shifting to this wider scope in understanding infectious disease relates to time and temporality – for process-based accounts of the gestation of infectious disease tend to miss the ruptures or radical changes that mark disease events. As Lynteris (2014) notes, for example, syndemics and other processual approaches to epidemics have a 'blind spot' in that they fail to acknowledge or take seriously the experiential and eventful nature of disease. They can suggest a rather seamless or linear temporality to disease that emerges from already established conditions. But disease is not experienced in this way. Instead there are bifurcations or radical shifts, particularly once an epidemic or illness takes hold. For Lynteris (2014: 26), infectious disease epidemics introduce 'a hitherto non-existent mode of being, a being which is at one and the same time pathological and infectious, as the central transformative factor of social life'. Pathological lives are punctuated then by social and biological events. These events in turn change the dynamics and social experience of epidemics and epizootics. They are not, as a result, all of a piece.

This is indeed the case, but on its own it does not take us far enough or indeed explain something of the eventness of disease. To do so there is also a need for space. In the following we therefore underline two key elements of our argument and that emerge from our analysis: the first concerns our spatial argument, while the second concerns the challenge to life politics that follow, we would argue, from this spatial imagination.

Time-Space and Intra-Actions

The event-ness of an infectious disease cannot just be a matter of time. Time in itself does not explain novelty, and radical change cannot be 'a mere rearrangement of what already is' (Massey, 1999: 272). For there to be time, for there to be ruptures or events, there is a requirement for there to be space. As Massey reminds us, for there to be change there is a need for mixture, heterogeneity and multiplicity – 'for there to be difference, for there to be time … at least a few things must be given at once' (1999: 274). The point for our purposes is that, certainly, disease is punctuated by tears in the fabric of lives, but these disruptions are also spatial productions. They are generated through spatial difference. Likewise, disease situations cannot simply involve the rearrangement of disease factors into an outbreak or illness in the lead up to turning pathogenic. Instead, and for there to be a disease situation, eventful juxtapositions are required. In analytical terms, pathological lives require an interrogation of multiple space-times, their intra-actions and interferences (Barad, 2007), in order that we can diagnose a disease situation.

In other words, in the accounts we have given here, infectious diseases are made through the intra-action of matters and processes whose own spaces and times are also not 'of a piece'. These are what we have referred to as borderlands, and are 'marked by the intersection of multiple spatio-temporal [dis]orders' (Sassen, 2006: 392). At various points in the book we have invoked the concept of borderlands to convey this sense of dynamic intra-action. This thickened space where profit margins meet the microbiological flora and fauna of guts and where wild bird habitats brush up against the practice of genetic sequencing is not just a matter of 'density'. As we noted in Chapter 3, these intra-actions are not simply about a coming together. Rather it is the folding together and stretching apart that matters in the changing dynamics of pathological lives. In other words, we need to attend to the spatialities or topologies that can produce the events that we call disease or illness.

Turning to topology is a way of bringing the temporal and spatial together, it is a way of understanding the condition that exists between continuity and disruption (Adam, 2004) – or the distortions or deformations that enable continuity under transformation (Chapter 3) (Allen, 2016). These twists can occur in the molecules that make up a virus, in the guts of chickens, in the bodies of pigs and in the supply and value chains that generate safe food. Time in this topological sense is not a matter of simple process, or indeed something punctuated by events, but is characterised or produced through a multiplicity of materials, processes and logics. A useful image here is the swirling waters that Steven Connor uses to capture Michel Serres's understanding of topology. Connor (2004: 109) writes that in 'place of a line of history, [we have] a series of different figurings of time, based on dynamic volumes.' These volumes are, he says, like a river, they fork, branch, slew, slow, roll back on itself and so on. Matters then do not simply transform in time, or shift from one state to another through time. Rather, spatial topologies *release* time.

As a corollary, multiple spatial-temporal disorders release pathologies. They make disease events possible. Life becomes pathological not only through disruption or rupture but through the flows and intra-actions that can deform and distort life. Importantly, this is not a linear process, nor is it simply about density. It is instead a matter of understanding the stretching and intensities that can accrue within disease situations.

A crucial distinguishing feature of our argument (compared to some process accounts of disease) is that we are not simply adding matters together to reach a disease crisis. This is more than a list of disease factors. We are instead teasing out the composition of multiple space-times, the intra-actions that change the dynamics of the situation. Situations are not in that sense accounts of everything, they are not systems or structures; they are relational accomplishments that are provisional and subject to change.

This is an empirically specific issue – as we have shown. The mixing together of concentrated animal feeding operations with casualised labour and fragile chicken endocrine systems produces an intensive coupling that can lead to food-borne disease and to the rapidly changing ecologies of viral quasi-species (Chapters 3 and 4). Elsewhere, the decline in resources necessary for food regulation (Chapter 6) as well as a reduction in the expertise necessary to patch together healthy lives (Chapter 5) suggest that pathological lives are borne out of being stretched in other ways. The antidote to this diagnosis is, as we have seen in many of the chapters, an ability to attend to multiple spatio-temporal tensions. Biosecurity becomes less about keeping matters out, and more a matter of learning to 'know with' a multifarious cast of others (Chapters 5–8).

Indeed, for there to be change, for there to be good biosecurity and good life, there is a requirement to note how particular practitioners can manage the tensions of that situation. Only those who know birds, viruses, meat, businesses, or in the words of Stengers (2009: 38), those who experience the 'bewildering variety of what it means to be both in touch with and touched by "reality"' can remain faithful to and empower a disease situation. Our concern, as we have stated throughout, is that those minority practices are themselves under threat in the name of greater biosecurity.

The question becomes, how to compose life differently, in ways that are less pathological? This is our second issue and concerns the prevailing understandings of the politics of life.

A livelier Politics of Life

A politics of infectious disease requires that we interrogate *how* lives are put together or composed. One resource for doing this is biopolitics. As we have detailed in a number of chapters, biopolitics alerts us to the ways in which life has become intricate with politics. Life is therefore *of* politics, becoming defined through and altered as populations and bodies become subjects of governance.

As an analytic, biopolitics tends to emphasise the gradual assumption of a power *over* life through its continuous regulation and control. Yet, as we have demonstrated, this control is always likely to be imperfect, or liable to failure, *and* is structurally bound to defeat itself. If to live and to flourish is to circulate (and move), then a prophylactic approach to security can only produce stasis and death. Security in this sense cannot be about the *protection* of life, but is concerned with 'securing the contingent freedom of circulation' (Dillon and Lobo-Guerrero, 2008: 282). Here, the powers *of* life and its flourishing become part and parcel of its optimisation.

One particularly apposite rendering of this tension for understanding pathological lives is immunity – not as a mechanism used in defence of a self but as a relation between self and other that generates new possibilities for health. Taking the maxim that what does not kill you makes stronger, immunity has offered us an alternative to the powers-over-life. Immunological intra-actions of molecules, microbes and bodies has provided an opening into a life politics that understands self and non-self as in relation, and immunity as a co-production. Optimising life, in this reading, requires attending to the intrication of people, non-human animals and microbes (Chapters 5 and 8). There is a flattening, you could say, and a challenge to the biopolitical imaginary that tends to start from a pre-established distinction between proper (human) and improper (less than human) forms of life.

We have taken this as a first step in challenging any aim to optimise life that is based on pre-established notions of species identity, self, norms of life or health. Going further, the normal and the pathological are clearly not predefined or necessarily in opposition. Health is not something that can be defined once and for all, and is not a timeless ideal or norm from which there are measureable departures or deviations (Canguilhem, 1991 (1966)). Instead, we have followed a variety of authors (and to an extent those who practice health) who view pathology as a push that stretches the normative basis of functioning. Diseases in this view not only challenge states of being, they may bring new ways of life or counter-norms into being (Schillmeier, 2013). These counter-norms can unsettle any pre-set notions of proper or normal life. They provide openings for thinking again and for challenging the biopolitical.

If the biopolitical, in some hands at least, seems too wedded to a pre-established division between proper and improper life, or normatively focused upon a version of health and species being that is exclusive of pathology, then the alternative we have worked with here is the intricate constitution of life and pathology. The question becomes, though, how might this intricate composition of life lead to a re-oriented politics of life? And how can we avoid veering between an exclusive focus on human being or a general dispersal of all being (Chapter 8)?

Esposito's immunological treatise is, as we have suggested, promising as a means to affirm a politics of life, and to undermine the anthropologic that seems to inhabit certain versions of biopolitics. Nevertheless, in its adoption of the Spinozan 'principle of unlimited equivalence for every single form of life' (Esposito, 2008: 186) there is a danger that we are left with a neo-vitalist and impractical connectionism. Again we are faced, it seems, with a stark choice between life (all life) or the 'autoimmune disorder that is bound to eventuate if the continuum of life is broken' (Wolfe, 2013: 62).

Composing a politics of life clearly requires moving from such *principles* of 'unlimited equivalence' to the *practicalities* of living pathological lives. It is here that Stengers's cosmopolitics has offered us more resources. In a manner that shares something with both Canguilhem's and Serres's understanding of disease and illness as ushering in new modes of normativity, Stengers (2005a: 996) reminds us that cosmopolitics is concerned with the 'passing fright that scares self-assurance'. These frights are borne and nurtured not so much from a principle of unlimited equivalence but from the ecology of practices that itself needs to be nurtured if we are to be able to heed those passing frights and re-orient norms accordingly. Again, the notion of the disease situation, with its mix of diagrams and pathogenicities, materials, species and expert knowledge that are exposed by their obligations to more than human others, helps us to think again about self-assurance and the security model it assumes (or the shoring up of pre-established selves).

A disease situation is made, on our account, not only from the multiple space-times that twist and turn lives into more or less pathogenic states, but also from the ability or otherwise to register and use those twists as events. To do this requires all manner of practices – from knowing birds to knowing meat, from veterinary patching together of healthy lives on the farm to raising disease publics and working with viral quasi-species. Some of these practices are, as we have noted, under threat. In the shift to forms of surveillance that tend to undermine the hesitations and obligations that are part and parcel of observation, *we can lose the capability of allowing non-human others to force us to think.*

This is not, we have emphasised, a matter of assembling the stakeholders and governing with reference to the established norms. It is about allowing specialist practitioners, often working in a minor key, to carry on being forced to think as they engage with their more than human colleagues. It is about allowing those who would mark events, through their knowing of viruses, birds, pigs, bodies and so on, to generate new collectives that can remain faithful to such events. This is a minoritarian politics (Deleuze and Guattari, 1988). It involves a nurturing of different modes of existence in ways that allow deviation from the standard or norm.

A new Kind of Emergency?

The key shift here is a recognition that the practicalities and diagramming of disease as matters for exclusion from human spaces and or subjects of human control need to be countered by two developments. First there is the need to nurture a recognition of the limits of human agency and second to adapt living to what are multiple entanglements of human, non-human and microbiological beings.

Cosmopolitics for us engenders a responsiveness to non-human others and a willingness to use the passing fright that is emergent and re-emergent infections to re-frame the infectious disease emergency. This shift of attention to adequate responses provides us with a counter-narrative to disease control. Instead of policing life in terms of borders between health and disease, the alternative proposes the question of what would it mean to partially abandon strict bordering and invest instead in living pathological lives. 'Living with' in this sense becomes not so much an opening to all life, but a matter of working or labouring alongside others, of all kinds. Those others, from wild birds to wardens, domesticated pigs to food inspectors, viral swarms to virologists, are always composite – that is they form human–non-human ties that allow for an understanding of how life can be pieced together. It involves treating people from a variety of areas as spokes-people (Stengers, 2010b), as experts in their relationships and obligations to non-human colleagues.

It is a living with and an attempt to save from extinction those who appreciate the complexities of health and illness, the entanglements of pathological life, and, in doing so shift the debate from securing territories (be they national states, food premises, or bodies) towards building better health-care systems, improving animal and environmental health, reducing inequality and vulnerability, and stemming environmental and socio-ecological degradation. Some of this is about redirecting investment to basic care, some of it, as we have noted, is about recognising and valuing the skills that exist within food and farming systems, scientific practice and in our day-to-day enactments of public health.

Certainly, this redistribution of expertise may risk a return to the logic and politics of an ever-expanding regime of security, vesting new powers and responsibilities in non-state and sometimes reluctant actors as a means to displace accountability and costs, all under the auspices of an ontology (emergence) that has its own non-innocent history. As we have detailed in several chapters, these risks are evident. Nevertheless, the alternative proposition of working alongside those people for whom the microbes, pigs, cuts of meat, and so on force them to think is also made in the face of a tendency to simplify and codify life and living in the name of disease risk. Those spokespeople who live with and are animated by the complexities and nuances of human, animal and plant health, its multi-dimensional characteristics, are becoming endangered in highly capitalised and de-skilled food and health landscapes. This we regard as the real challenge to what should be a vibrant ecology of practices, or a distributed sensitivity to the multiple space-times of pathological lives.

There is then a need to arrest a tendency to oversimplify the food chain and downgrade agricultural, research and surveillance roles within the food, farming and health sectors. Indeed, the sources for making life safe may well be found in the maintenance of the difficult work that is involved in patching together healthy lives. Safe life depends upon a trans-disciplinary assemblage of expertise, from farmers to catching teams and slaughterhouse workers, from vets to virologists, from local authority health officers to wildlife volunteers and professionals. These people, who work with bodies and environments on a daily basis, and through their craft and non-human alignments, are forced to think and act upon the counter-norms of health and illness, need *not* be imagined as exemplary neoliberal subjects, nor as integrated and so simplified by a coherent health infrastructure. They are instead, the spokespeople of the human–non-human collectives that make safe life possible.

As a result, the emergency of emerging infectious disease changes. Rather than a matter of emergent life threatening the norms of human health, cordoned off in ever more sophisticated and policed spheres, emergent life generates the possibility for questioning norms, and for raising counter norms. The current emergency is not so much about emergence of infection, but of life that is deprived of qualities, lived as mere life, and lived at the biological threshold. Pathological lives are not then the problem, but part of the solution. They are the threats to self-assurance that can force us to think again.

References

Adam, B. 2004. *Time*. Cambridge: Polity Press.

Allen, J. 2016. *Topologies of Power: Beyond Terrritory and Networks*. London: Routledge.

Barad, K. 2007. *Meeting the Universe Halfway: Quantum Physics and the Entanglment of Matter and Meaning*. Durham, NC: Duke University Press.

Canguilhem, G. 1991 (1966). *The Normal and the Pathological*. New York: Zone Books.

Connor, S. 2004. Topologies: Michel Serres and the shapes of thought. *Anglistik*, 15.

Deleuze, G. & Guattari, F. 1988. *A Thousand Plateaus: Capitalism and Schizophrenia*. London: Athlone.

Dillon, M. & Lobo-Guerrero, L. 2008. Biopolitics of security in the 21st Century. *Review of International Studies*, 34, 265–292.

Esposito, R. 2008. *Bios: Biopolitics and Philosophy*. Minneapolis: University of Minnesota Press.

Foucault, M. 1973. *The Birth of the Clinic: An Archaeology of Medical Perception*. New York: Vintage.

Lynteris, C. 2014. The time of epidemics. *Cambridge Anthropology*, 32, 24–31.

Massey, D. 1999. Space-time, 'science' and the relationship between physical geography and human geography. *Transactions of the Insititue of British Geographers*, 24, 261–276.

Sassen, S. 2006. *Territory, Authority, Rights: From Medieval to Global Assemblages*, Princeton and Oxford: Princeton University Press.

Schillmeier, M. 2013. *Eventful Bodies: The Cosmopolitics of Illness*. Farnham, UK: Ashgate.

Singer, M. 2009. *Introduction to Syndemics: A Critical Systems Approach to Public and Community Health*. San Francisco, CA: Jossey-Bass.

Stengers, I. 2010a. *Cosmopolitics 1*. Minneapolis: University of Minnesota Press.

Stengers, I. 2009. Thinking with Deleuze and Whitehead: a double test. *In*: Robinson, K. (ed.), *Deleuze, Whitehead, Bergson: Rhizomatic Connections*. Basingstoke and New York: Palgrave Macmillan.

Stengers, I. 2010b. Including non-humans in political theory: Opening Pandora's box? *In*: Braun, B. & Whatmore, S. (eds), *Political Matter: Technoscience, Democracy, and Public Life*. Minneapolis: University of Minnestoa Press.

Wolfe, C. 2013. *Before the Law: Humans and other animals in a biopolitical frame*. Chicago and London: The University of Chicago Press.

Index

Pathological Lives: Disease, Space and Biopolitics, First Edition. Steve Hinchliffe, Nick Bingham, John Allen and Simon Carter.
© 2017 John Wiley & Sons, Ltd. Published 2017 by John Wiley & Sons, Ltd.